Kinabalu

The Haunted Mountain of Borneo

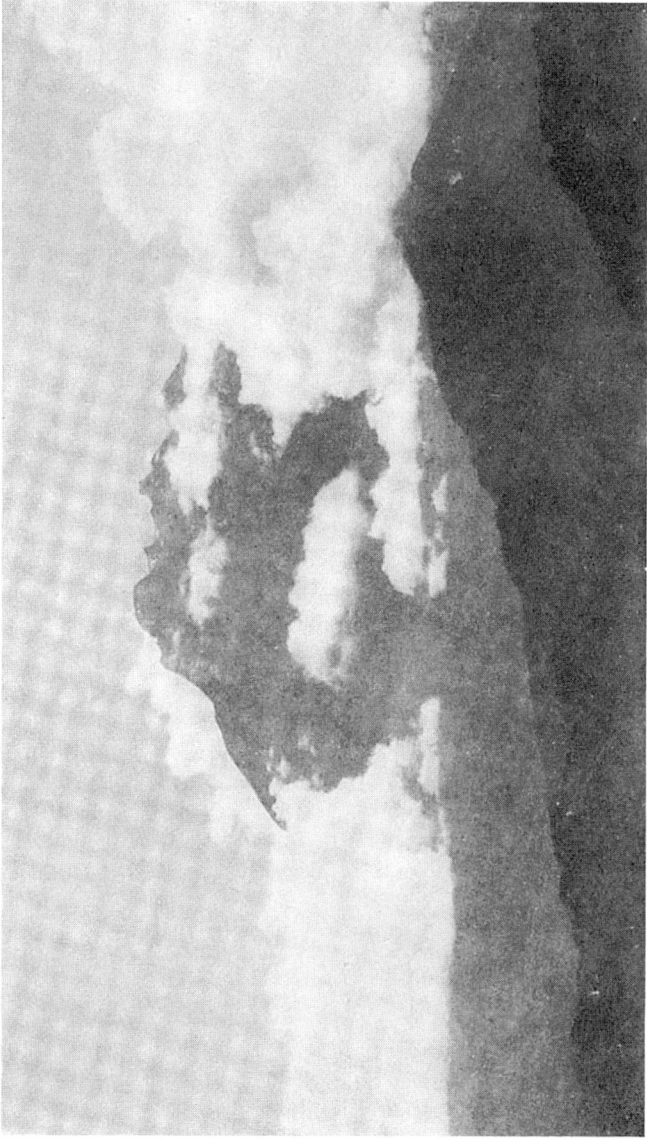

MOUNT KINABALU.

Kinabalu

The Haunted Mountain of Borneo

An Account of its Ascent, its people, Flora and Fauna

MAJOR C.M. ENRIQUEZ, F.R.G.S

With an introduction by
K.M. Wong

Opus Publications
Kota Kinabalu
2008

Published by

Opus Publications Sdn. Bhd. (183100-X)
  A913, 9th Floor, Wisma Merdeka Phase 1
  P.O. Box 15566
  88864 Kota Kinabalu, Sabah, Malaysia
  Tel: 088-233098 Fax: 088-240768
  e-mail: info@nhpborneo.com

Kinabalu—The Haunted Mountain of Borneo
An Account of its Ascent, its People, Flora and Fauna
by C.M. Enriquez
With an Introduction by K.M. Wong

First published 1927 by H.F. & G. Witherby, London.
Reprinted 1995 by Natural History Publications (Borneo) Sdn Bhd
This impression published in September 2008
by Opus Publications Sdn Bhd, Kota Kinabalu

Introduction copyright ©2008 Opus Publications Sdn Bhd

Cover photograph copyright ©2008 Robert New

ISBN 978-983-3987-25-2

Printed in Malaysia.

Contents

Introduction vii

Preface xi

Chapter I   The Kinabalu Expedition 1

     II   The approach-Kotabalud to Kabayan 5

     III   The approach-Kabayau to Dallas 17

     IV   The Ascent 27

     V   The Summit 39

     VI   Bundu Tuan 51

     VII   Benoni 55

     VIII   Interior 59

     IX   Indigenous races of British North Borneo 65

Conclusion 73

In Memoriam 75

Appendix I   Itinerary 79

     II   Botany of Mount Kinabalu 81

     III   Spiders of Kinabalu 85

     IV   Birds obtained by the Kinabalu Expedition 87

     V   Distribution of Butterflies
                 in British North Borneo 93

     VI   Supplementary list of butterflies obtained
                 and identified in British North Borneo,
                 and not mentioned either in the text
                 or in the index for butterflies 95

INTRODUCTION

Enriquez's short account, *Kinabalu—The Haunted Mountain Of Borneo* requires only a brief introduction in that it is a succinctly composed record of an expedition to Borneo in 1925, made principally to collect zoological specimens. This was when Borneo was a far-away land of *orang-utan* and headhunters, savagery and piracy and when (under the British North Borneo Company) "the question of communications remains... in the air." It is still a far-away land of *orang-utan* for many Europeans, but the headhunting and the savagery has gone, remaining only in the annals of the past.

The reader can finish the book in one or two days at the most, but the admirer of Kinabalu and keen collector of Borneo literature will notice that some of the most beautifully written passages on Kinabalu that portray its starkness and many wonderful moods exist in Enriquez's book. Even in those days, at a time when it was thought there had been only 25 documented visits and sacrificial rites were an essential condition to climb the mountain, he notes that "Kinabalu is a sort of Mecca for naturalists." Enriquez was a military man and a sort of naturalist, knowledgeable about butterflies and birds and, although not a resident of North Borneo, makes some rather sharp observations on natural history aspects of the mountain. He comes across as a curious visitor enchanted with Kinabalu, and in such a state prone to forgetting himself, although understandably so. He declares his book seeks to "avoid... the deplorable terminology beloved of scientists" but (even as he refers to spiders as those "unpleasant insects") engages in mentioning suites of latin names of butterflies and birds, which otherwise are charmingly described. Some of his observations on the races and customs of North Borneo may appear coloured and unacceptable, as can understandably be made by a member of any administrating party over their colonial charges (and, indeed, some of the most unusual stories could be uncritical repetition of what has been related to the apprehensive traveller), but Enriquez displays his fondness for the gentle aspects of local communities in the Far East, not least by dedicating his book to his Kachin companion from Burma, Sau Nan, an associate from the British Indian army, and by declaring, in connection with his North

Borneo experience, that "superstitions... denote a sense of humiliation in the presence of great forces where perhaps more arrogant people betray a profounder ignorance by ridiculing them." This is a wonderful mixture of attempts at accurate natural history observations and calm expressions of what has been today called culture shock.

These are the beginnings of the state of British North Borneo, then with a population of only 357,000, when the journey on foot from the town of Kota Belud to Bundu Tuhan village, where the main assault on Kinabalu began, took more than a day (nowadays less than an hour by car), to the extent that Enriquez observed was "not to be recommended generally for women."

Readers can now sit back and ponder over the misguided predictions of Enriquez, who suggested himself as a "cynical Twentieth Century being." In 1975, fifty years following Enriquez's expedition, 2126 persons made documented climbs to the summit of Kinabalu, and nearly twenty years later, in 1994, 29,574 persons had done so. Enriquez had remarked that "it cannot be supposed that when British North Borneo is more opened up its goverment will take so personal an interest in travellers" and that "in the future this wonderful mountain will be even less frequently visited than in the past." Perhaps it was this uncertainty that such an amazing creation of nature might ever be enjoyed by most people, that has prompted some of the most beautifully written expedition accounts in the past. But even as Enriquez downplays most other aspects in "Sarawak, with its White Raja—and British North Borneo, the property of a company—whose conception furnishes the history of the Malay Archipelago with two romantic passages," he was entirely right in revelling in the mystery of Kinabalu and its forests where, he wrote, "each bend... did reveal some new and unexpected wonder."

K.M. Wong
Forest Research Centre,
Sandakan, Sabah

September, 1995

*This book
is affectionately dedicated to*

SAU NAN

*a kachin of Lahtaw Hpahkum, Bhamo, Burma,
and for many years
my devoted and trusted friend*

PREFACE

It may not be out of place to begin an account of Mount Kinabalu with a brief sketch of the history of British North Borneo. This will savour of the Snark Hunter's preface, which started with 'his dear Uncle': but it is open to impatient readers to follow the Bellman's example and 'skip the dear Uncle.' On the other hand there are those who prefer to have light before venturing into dark places: and it is possible (present company excepted) that there may even be a few whose education in Bornean affairs has been woefully neglected. Borneo is the second largest island in the world, with an area of 286,000 square miles, but a population of only 1,700,000. This population is not related to Orang-utans—at least, not more closely than the rest of us: but at no very distant period the people of Borneo had some unfortunate habits, such as, piracy and headhunting, for which, in spite of the able writings of many travellers, they are still better known than for their domestic virtues. It is almost disappointing to find them in prosaic fact, quite a mild and pleasant folk, and their country extremely beautiful.

Borneo has been known to the Chinese from very early times, and Chinese coins dating from B.C. 600 to 102, and from A.D. 588 onwards, have been found in Sarawak. It is probable that Genghis Khan extended his conquest of China (1206–27 A.D.) to Borneo, where Chinese had already settled: that Kublai Khan, thus brought into conflict with the Javan Empire of Majapahit, invaded the island with considerable forces in 1292: and that a Chinese colony was subsequently established in North Borneo. It is certain that in the fourteenth century the Sultan of Brunei, once a vassal of Java, transferred his allegiance to China. There are fragments of evidence to show that the present comparatively savage races of the Interior were at some forgotten period in touch with Chinese civilization, as well as with the Hinduism of Majapahit. The Dayaks retain traces of contact with Java, and the Dusun of even more intimate relations with China.

From Sumatra and the Malay Peninsula, Malays spread to Borneo, their colonies gradually being consolidated along the coast into Sultanates like those of Brunei, Sambas, Banjermasin, Koti, Pasir, Tanjong and

Pontianak. In the fourteenth century the Sultan of Brunei was a vassal of the Javanese Empire, but in 1370, as we have seen, transferred his allegiance to China.

The first Europeans to visit Borneo were the Spanish when Magellan circumnavigated the World in 1521. Brunei was then a prosperous state with a large Chinese population that disappeared in the troubled times that followed. From 1530, the Portuguese maintained a regular intercourse between Borneo and their colony at Malacca, until expelled from there by the Dutch in 1641.

The Dutch connection dates from 1600. Factories were planted at Pontianak in 1778, but withdrawn in 1791. Then in 1818, two years after the restoration of Java by the British, these factories were recommissioned, and from that period the present Dutch Residency of Western Borneo dates. In the meantime, as the power and authority of the Malay Sultans decreased, piracy in Bornean seas grew and reached really frightful proportions by the middle of the nineteenth century. Intercourse with China ceased, and trade was practically killed by the combination of this lawlessness and the Dutch system of 'Monopoly.' As piracy increased, it was fostered by Malay Chiefs who strove thus to revive their fortunes. Slavery was rampant. The coast people of Borneo were carried off wholesale. Many fled further into the Interior, land went out of cultivation, and Borneo gradually sank into a hopeless anarchy and misery.

To this wrecked society in Sarawak came James Brooke in 1839. We cannot here relate in detail how he championed the poor Dyaks, reduced Brunei to a petty state, subdued piracy, and acquired independent sovereign rights in Sarawak. It is a long and romantic story, but all these things happened and are told in my *Legend of Malay*, and in 1888 British protection, long and vacillatingly withheld, was formally extended to Sarawak and its English Raja.

In the meantime another romantic passage of history was developing in what is now British North Borneo, and it begins with William Clarke Cowie who sailed East from Glasgow in a 14-ton yacht to found a kingdom in 1872. His success in gun running, in defiance of Spanish authority in the Philippines, so enchanted the Sultan of Sulu (who loathed Spaniards) that the Sultan gave Cowie sovereign rights over North Borneo, in the lighthearted way they had in those days, that he might establish a convenient gun running base in Sandakan Bay. All would have been well, had not a Mr Torrey, styling himself 'Raja of Marudu and Ambong,' claimed that the country was his. It now appeared that the Sultan of Brunei had given Torrey the same rights over North Borneo as Cowie had received from the Sultan of Sulu. To cut a long story short, it will suffice to say that the parties came to an agreement. The possibilities of exploiting the country were obvious: the British authorities, who had acquired Labuan from the Sultan of Brunei, favoured the project, and there were financiers in the background—notably Dent Brothers. As a result, sovereign rights were formally acquired from the Sultan of Sulu in 1878, just six months before the Spaniards ate him up.

The willingness of the natives of North Borneo to accept the transfer is quite remarkable. There have never been any operations that deserve to be classed as more than police expeditions.

The new owners of North Borneo appointed agents, issued trade rules, and cultivated friendly relations: but they were in no position to do more. Money had to be found, and protection. A 'Limited Provisional Association' was formed, taking unto itself experienced and influential men like Sir Rutherford Alcock and Admiral Sir Henry Keppel of Sarawak fame. In 1878 a Royal Charter was asked for, and after a hot controversy, and violent protests from the Dutch and Spanish (neither of whom had the ghost of a claim), the Charter was granted in 1881. In 1888, North Borneo, together with Brunei and Sarawak, became a British Protectorate.

The British North Borneo Company now organised its Government on a permanent footing, deriving its revenues rather from the administration of the country than from trade. British North Borneo was thrown open to private enterprise, and the Chinese were encouraged to immigrate—since more population was the first crying need. And it is yet a crying need. Free passages are still given to Chinese immigrants, who would avail themselves readily of such concessions but for the veiled hostility of the authorities in China. In 1883, tobacco was first planted. The boom of 1885 eased the Company's finances, but tobacco is not extensively cultivated now.

The district of Api Api, on which now stands the town of Jesselton, was purchased from Brunei in 1898, and the frontiers were generally rounded off to march with those of Sarawak and Dutch Borneo. Such railways as there are were constructed between 1896 and 1905. They, however, run nothing better than goods-passenger trains, and have been a heavy expense. The construction of roads is also costly, and the question of communications remains at present in the air. The projected 'Grand Trunk Road,' starting boldly for some twenty miles at both ends, has there halted indefinitely. Motor roads are an urgent need, but the present thinness of the population does not seem to justify the expense. The question is, would roads, if made, attract a population?

The country, as far as is known, has few metals, and has never had a mining tradition. It is essentially agricultural. It possesses valuable forests, rich, fertile uplands, and a comfortable climate. Rubber, first planted in 1893, has had a successful history.

There can be no doubt but that British North Borneo has a future before it. But, as in the beginning, so now—it needs money and men. The population is only 357,000, or about ten to the square mile. That is to say, British North Borneo, as a whole, is to be compared with the most empty spaces of Burma and the Malay Peninsula, where, in Möng Pan and Pahang, the population is seven to the square mile.

C.M. Enriquez

MAP OF
BRITISH NORTH BORNEO
To Illustrate

"KINABALU-THE HAUNTED MOUNTAIN OF BORNEO"

— ROUTE DESCRIBED

SANDAKAN

SANDAKAN RESIDENCY

INTERIOR RESIDENCY

COAST RESIDENCY

MARUDU BAY

KUDAT

MT KINABALU

RANAU

KOTABELUD
PENEWELAN
TAMU DARAT
KABAYAU
KAUNG
DALLAS
BUNDU TUHAN
KINOHOK

TAMBUNAN

TUARAN

JESSELTON

USUKAN BAY
TEMPASSUK RIVER

BRIDLE PATH

KENINGAU

KELALAN

ENOM

BEAUFORT

EDGE OF PADAS

PAPAR

BENONI

KIMANIS BAY

THE KINABALU EXPEDITION

Kinabalu (13,455 feet) is a mountain rising up out of British North Borneo. It is the highest Peak in the Malay Archipelago, and by reason of the specialized forms of its flora and fauna is a sort of Mecca for naturalists. Since Captain Sir Edward Belcher, R.N., and Sir Hugh Low explored it in 1814 and 1851, it has been climbed and visited by Spencer St John, author of *Life in the Forests of the Far East* (1858); Giordano (1873); Burbidge, author of *Gardens of the Sun* (1877); Peltzer (1879); Little (1887); Whitehead, who wrote a standard work *Exploration of Kina Balu* (1893); Haviland (1892); Hanitsch, Burls and Waterstradt (1899); Pilz, Foxworthy, Learmonth and Bunbury, Maxwell and Miss Gibbs (1910); Moulton (1913); Mrs Swinnerton (1915); Evans and Sarel (1924); and myself (1925). Captain Learmonth, R.N., made a survey of the mountain,· and the heights I have given are taken from his map. Moulton has written a useful summary of the previous expeditions. At the summit I found a bottle containing what is said to be the record of an ascent by two Chinamen, O Hau Teck, a shop-keeper of Jesselton, and Leong Si Toon, Secretary to the Chinese Consul in British North Borneo. The number of persons who have visited either the summit or the higher altitudes would thus appear to be twenty-five, apart from any who were attached to these various expeditions, but whose names have been forgotten. Since the discovery of the route now generally used, there is no danger or special difficulty in the ascent. But, rising as it does, some day's march inland from the Bornean coast, Kinabalu still possesses a certain romance, for it has jealously guarded its mysteries from all but a select band of pioneers.

Having two months' leave due while stationed at Taiping in the Federated Malay States, I organized the Expedition of 1925, which is here described. The primary object of the journey was to collect butterflies, but as I approached the majestic and awe-inspiring battlements of Kinabalu, I became more and more obsessed with a

1

longing to reach the summit. A peak so little known, and with an appearance so formidable, yet (as I knew) comparatively easy of ascent, proved irresistibly alluring. I felt that to climb it would set the mark of success on our research; for besides butterflies, which are my own hobby, I was entrusted with several commissions. For Mr Abraham of Taiping, an authority on spiders, I was to collect those unpleasant insects; and I must confess that, though I know nothing whatever about spiders, the number and diversity of types obtained at the base of Mt. Kinabalu was altogether astonishing. By Mr R.E. Holttum I was requested to collect ferns and mosses for the Singapore Botanical Gardens; and Mr Chasen, Director of Raffles' Museum, attached to the Expedition a Vertebrate Collector—a young Seribas Dayak, called Mingga. I am indebted to Mr Chasen for the most patient identification of the collection. The Expedition had to a certain extent a scientific aspect; but in describing what we saw I have here used the simplest language, avoiding as far as possible the deplorable terminology beloved of scientists.

My party consisted besides of Rifleman Bilu Gam, 2/20 Burma Rifles, and Sau Nan, a trusted friend of long-standing, who has been my right hand man in many ventures. Both these are Kachin hill men from the North-East Burmese Frontier, experts in jungle lore. The wisdom of this selection was all the time evident on Kinabalu, where Sau Nan's resource and energy in rain and cold were invaluable. There was also my Burmese servant, Maung Ba Kye, and a locally engaged Dusun called Paud to act as interpreter.

Leaving Singapore on the 31st of May (1925), we spent a day at Miri in Sarawak, and approached Labuan Island on the early morning of the 4th June. Here we had our first distant view of Mt. Kinabalu away across the sea, its high, level summit upraised against the dawn. Very remote and aloof it seemed.

Labuan is a little, flat island in the Bay of Brunei. Its importance, now that coal has failed, lies in the fact that it is a despatching depot for Brunei and several districts of British North Borneo. Both countries export much sago—the latter shipping 90,000 bags annually. Incidentally, I noted with interest the manufacture of sago. The nasty pap that is hopefully fed to tender infants is prepared in stinking mangrove water that has first been used as a latrine.

On the whole, Labuan is as depressing as its sagos. There is a small but safe harbour—and, indeed, the Malay name *'Labu-an'* means 'Anchorage.' Behind the wharf is a street of Chinese shops, and a narrow beach fringed with Casuarina trees. The harbour extends into a sort of Mangrove lake, with one or two Bajau villages built on posts out of the sea; but all these interests are exhausted in ten minutes. The real significance of the place is that it is British territory—a sentry, as it were, to protect those two States —Sarawak, with its White Raja—and British North Borneo, the property of a company—whose conception furnishes the history of the Malay Archipelago with two romantic passages. Labuan was acquired from Brunei in 1845; and since 1906 has been part of British Malaya for purposes of administration.

On leaving Labuan, the uplands of Brunei are seen to be considerable, and these ranges, which must rise to 5000 feet, extend into British North Borneo. The coast becomes more and more attractive. Jesselton, which we reached the same evening (4th June 1925), is sheltered by some islands. The town originally occupied Gaya Island, but was shifted first to Gantian *('Place of Change'),* and finally to its present site on the mainland. The residential quarter is built on a low coastal ridge, behind which are higher mountains culminating, as everything culminates in North Borneo, in Kinabalu.

Mr G. C. Irving, the Resident, came aboard to meet me. And here I must acknowledge, if that be possible, the kindness, encouragement and hospitality that was extended to me by all officers of the British North Borneo Company's Government. Mr Fraser, the Governor, dined me; Mr Irving, the Resident, put me up in his house, planned, organized and supervised every move; while later, under his directions, Mr D. C. Round-Turner, the District Officer of the area concerned, travelled with me to the very base of the mountain, and did not leave me till all matters relating to guides, porters and routes had been arranged. I felt at the time that it was quite impossible to express my gratitude and appreciation for all this attention; and though I shall not refer to it again, it must be remembered that Mr Irving's patient and efficient help was always behind me during the ascent of Kinabalu. I should also like to take this

3

opportunity of acknowledging the generous contributions made by the British North Borneo (Chartered) Company towards the cost of publishing this book.

It cannot be supposed that when British North Borneo is better opened up its government will take so personal an interest in travellers. At the same time, the Dusun inhabitants round Kinabalu are extremely simple folk. It is improbable that they will advance as quickly as the more civilized communities. And since their assistance in the matter of Kinabalu is essential, one feels that in the future this wonderful mountain will be even less frequently visited than in the past.

For the jagged heights of Kinabalu are the mysterious retreat of the Dusun dead, where mortal man does not care to intrude upon that cold and bitter sanctuary of ghosts.

Chapter II

THE APPROACH—KOTABALUD
TO KABAYAU

An ascent of Kinabalu requires a good deal of preparation. In Singapore, Raffles' Museum supplied traps, spirits, test-tubes, specimen boxes and a collector's gun; and on arrival at Jesselton, maps, books and stores were added. Decisions had to be made with regard to routes, and while these arrangements were in progress the Expedition remained at Benoni, a sea-side village two hours by rail up the coast. Here I took a house, and Benoni was our *pied-à-terre*. In the meantime stores were obtained from the ever-courteous and reliable Chinese firm of Ban Guan, and packed in *'Bongons.'* The *'Bongon'* is a jar-shaped basket made of bamboo, bark and wood; very light, completely waterproof, and specially suited to the taste of Dusun porters who carry them on their backs, and indeed decline to lift loads in any other form. If the *'Bongon'* has a fault, it is that outwardly all *'Bongons'* are alike: while inwardly all assume a uniform confusion. If a tin of soup is required, there are inevitably layers of shirts, rat-traps and potatoes to be excavated. The most careful repacking will not for long keep the *'Bongons'* clear: for should the onions be extricated from the bedding, that onion *'Bongon'* will weigh about a ton, and will become an object of scorn and derision to Dusun maidens, whose height is the proverbial 'two jam-pots.'

A steamer leaves Jesselton for Usukan Bay once a fortnight: and on the 10th June, Mr Irving, with his usual kindness, saw the Expedition embark. I need hardly add that these little attentions from the Resident are worth a great deal to a traveller in the East, where the attitude of superior officers is reflected all along the line down to the humblest Head-man.

The sea was calm, but Kinabalu was lost in brooding clouds. The voyage occupies only four hours, and we reached Usukan Bay in the afternoon. It is a pretty little land-locked cove, with a pier; and though there is no village, the arrival of the steamer is a social event, and a little crowd of

Dusun and Bajau had collected. With them was Mr D.C. Round-Turner, District Officer of Kotabalud, who had come to escort me to the base of the mountain.

Why Round-Turner was not permitted to make the ascent with me, as he much desired to do, I cannot say. Yet such is Government's attitude with regard to its officers and Kinabalu. The result is that, with a few exceptions, the only people who climb the mountain, and who profit by all the experience that that entails, are those who are not residents of Borneo.

On landing at Usukan, we crossed a low ridge, ferried a river, and then set out on a seven mile ride along a bridle-path that runs at first through swamps, then over laterite hills, and so descends to the great plain of Kotabalud. Kotabalud is the head-quarters of the District: a small town of forty or fifty Chinese shops, with Bajau houses scattered about, and Round-Turner's hospitable little house on a hill that rises out of the plain. Above all towers Kinabalu, flushed marvellously pink in the sunset. A few pigeons passed over as we rode. The little black-and-gold Iora, now in full breeding plumage, was plentiful; and a certain thrush has quite a remarkably sweet song. On the whole, birds were more plentiful than I expected. A lovely blue-green Kingfisher, *Halcyon chloris cyanescens*, with a white breast and collar, kept watch over the marshes. (One of these birds flew on board the steamer off the Sarawak coast before we had sighted land.) A nervous little Fan-tailed Warbler that is common on the plain was fussily preparing to roost, and as darkness fell, Night-jars rocketed up from the path.

Sunrise on the following morning gave us our first really good view of Kinabalu, which from Kotabalud is seen broadside on high above everything else, and sweeping up into the sky in mighty precipices. We had not seen it properly again since the distant view from Labuan, and its uncompromising splendour, now revealed at comparatively close quarters, was quite awe-inspiring. I do not hesitate to admit that it filled me with misgivings.

After seeing Kinabalu from many angles, I think this aspect of it from a distance of, say, 25 miles, is probably the most imposing of all, displaying as it does the full majesty and aloofness of the mountain. Its

6

relation to the surrounding country, its superb loneliness, are here seen in proper perspective.

We halted a day in Kotabalud to procure porters and to repack the 'Bongons' which were already in confusion. It was after this that I abandoned as a hopeless waste of time any further efforts to keep them tidy. The Expedition had twenty light loads, which seemed to be moderate; and even so, half the baggage was naturalists' paraphernalia. Round-Turner also thought we might get butterflies at Kotabalud, but the Lepidoptera was entirely disappointing, and limited to a few miserable *Terias* and two *Junonia—wallacei* and *atlites*. I did, however, obtain *Delias pandemia* and *Hypolymnas misippus* (Plate I, 18), the latter a black butterfly with two large heliotrope spots on the fore-wing and one on the hind-wing. The reverse shows these heliotrope marks on a light-brown ground, and has a finely etched marginal pattern in black and white. This is the male. The female is entirely different, being a mimic of *Danais chrysippus*.

The name 'Kotabalud' means 'Hill Fort.' Bajau and Ilanun occupy the plain, and Dusun the hills round about. The Bajau seem to be rather an unpleasant crowd. Government keeps a herd of cows and a few tame deer in Kotabalud, and during my stay, either for vendetta or by way of a joke, some of the cows were stabbed and one of the deer was killed.

On the 12th June we started off again towards the mountain, following more or less the left bank of the Tempassuk River, which has its birth amongst the cataracts of Kinabalu. The hills through which we passed are at first rather bare, but later on the forest improves, and there are promising patches of butterfly jungle near several little streams. The path we followed is a mere grass-grown bridle-track, but it leads direct to Tambunan and Tenom in the far interior, with little bamboo rest-houses at every stage. The views of Kinabalu in the morning sunshine, and later when clouds piled up round it, with the Tempassuk River brawling in the foreground, were very beautiful. The country is fair butterfly ground, but I was disappointed to find the types consistently Malayan all the way to Tamu Darat. *Pathenos silvia lilacinus* (Plate II, I6) flitted through the foliage. *Hebomoia glaucippe* (Plate I, 7), the great orange-tipped white, dashed about at high speed. The female, of which one specimen was obtained, is white with blackish tips, and rusty underparts. It is

7

considered a rare trophy by collectors. Two brown Danais, *plexippus intensa* and *chrysippus chrysippus linn*, were seen, as well as the beautiful red *Cethosia* (Plate I, 20), and *Cynthia erota erotella*, the last being met with all the way to Dallas. The common Appias is *Lyncida enarete*, which is later mixed with *Huphina hespera* (Plate I, 22), a White with a touch of rich light-orange on the hindwing, until finally both give place near Tinompok to the curious black-edged *Appias pandione whiteheadi* (Plate I, 32) which is the form peculiar to Kinabalu. A *Papilio* with slender tails, *Antiphates itam puti* (Plate I, 25), also occurs here, though only one was seen on the whole journey. It gives place on Kinabalu to *Papilio stratiotes*, which has a red patch near the tail. *P. stratiotes* is figured in Whitehead's book, but unfortunately I never found it. However, an equally curious transition is that of *Ragadia crisia siponta*, a little striped grey butterfly with a marginal row of shining silver dots on the reverse, which occurs all the way up, till, beyond Dallas, it is replaced by a whitish, black-edged form, *Ragadia melindera annulata*, which is peculiar to Kinabalu. These are some of the most remarkable transitions; but all the way up, as the altitude gently increases, the old types fall out and new take their place, till between Dallas and Tinompok, on the actual base of the mountain, the peculiar forms reach their maximum. Down here, on the early marches, the common Papilios are *Sarpedon, Jason, Memnon sericapus* (Plate II, 5), green *Agamemnon*, black *Polytes theseus* and *Aristolochiae antiphus*; and two large, black swallow-tails, one of which has white or creamy plates, and is *Nephelus albolineatus* (Plate II, 10). The other has rusty marks on both wings, and though it has also been identified as *N. albolineatus*, I am not at all sure about it. Both these swallow-tails inhabit jungle, and dart along erratically and at high speed through the thickets. At this low level occurs the great black-and-yellow Ornithoptera *Amphrysus flavicollis* (Plate II, 2 and 3). This species, which seems to be the commonest in North Borneo, resembles, especially in the male, *ruficollis* of the Malay Peninsula, except that there is no scarlet on the body, but yellow, as, I believe, is the case with Javan forms. It is rather curious that this commonest Bornean Ornithoptera should be nearer to the Javan than to the Malay Peninsula form, considering the ancient isolation of Java from the surrounding countries. In the female (Plate II, 3) the black marginal spots of the hind-wing are elongated and run together, while the veins of the fore-wing are broadly edged with yellowish-white, giving this great butterfly a much paler and handsomer appearance than *ruficollis*. At a slightly higher altitude—say about 1500 feet—this Ornithoptera disappears and is replaced by *Brookiana* (Plate II, 1).

Generally speaking, the Lepidoptera of North Borneo is therefore Malayan in type; yet the variations are sufficient to constitute a difference in species in many cases. The absence of certain common Peninsula papilios was striking, though it is not suggested that they do not occur. Still, amongst the profusion of butterflies found in this rich field, none of the papilios *Delesertii, Doubledayi, Iswara, Macareus* or *Neptunis* were seen; nor am I certain that I met *Nox*.

With regard to insects at this low elevation, I got a few Fulgoridae, and a fine dark green beetle of the family Buprestidae, *Chrysochroa opulenta* (Plate II, 25), with a narrow white bar on the elytra. The wings are opaque and greenish, and when spread expose a shining green thorax and an abdomen of which the middle four sections are white. The Beetle is two inches long, and altogether very handsome.

Tamu Darat (the name means 'Up-country Market') is a long, grassy plain with green hills round about, and a little rest-house beside the river. A fair takes place every five or six days, but few villages are visible, and one begins to realize how very sparsely North Borneo is populated. On arrival we had a delicious bathe in the river; and after lunch the Dusun put up a deer drive, though without success. Not the least interesting features of the beat were the little Dusun dogs that yelped and shrieked in the jungle. These noises they can make, but it seems they are unable to bark. In spite of their small and mean appearance, the foxy little Dusun dogs have plenty of character. I watched several swim the river after their masters. The stream was swift, and it was evident from their yelps of distress that it required considerable courage to face it.

Though we saw no deer then, we got a Mouse Deer at the next stage. It was 24 inches long, and 12 inches high at the shoulder—a most dainty little creature. The small *Punai* pigeon is common; and at several places we got the huge *Pergam*, which seems to be a kind of green Imperial pigeon. These great birds are 17 inches long. The head, neck, throat and breast are light-grey. The wings and tail are dark-grey, with a green or bronze lustre according to the light. The under-tail covers are dull claret.

The Argus and Fireback Pheasant both occur in the country. Though the birds are rarely seen, the 'dancing floors' of the Argus Pheasant are sometimes found: and the natives value the splendid 'eyed' feathers, which are indeed beyond compare. Snipe and Golden Plover are plentiful in January.

Two Chiefs came into camp at Tamu Darat—one Keruak—a Bajau, a fine old fellow, enormously fat, and with an uncontrollably winking eye. This mountain of good humoured flesh rode a rat of a pony, carried a ferocious spear, and organized the deer drive. The other chief, Lumandan, was a Dusun of Kaung Village—a broken reed, apt to go sick when wanted, and by no means a fair example of his people, who are on the whole very attractive and engaging. He spoke of the necessity for offerings and sacrifices to the Spirits of Kinabalu, and that, as far as I am aware, was the extent of his usefulness. The sacrifices ultimately agreed upon were seven eggs and seven chickens, of which the total value was 77 cents. But the careful insistence on these paltry offerings, and the lengthy negotiations they entail, only tend to prove how real is their importance in Dusun eyes. The mountain is inhabited by the spirits of their dead: and though their conception of these Beings, and of Animism generally, is exceedingly confused, they would certainly never dare to make the ascent without performing the correct rites.

From Tamu Darat we marched 11 miles to Kabayau on the 13th June. We were travelling by easy stages to get fit after board ship, and gradually to harden the feet. Partly on this account, I declined the kind offer of riding ponies; but my chief reason for walking was that on foot one sees so much more of the country and its fauna. Our marches were short to admit of rambling, and it repaid us a hundred fold to travel thus, and to work over the ground a second time on our return: for the country from Tamu Darat to Kabayau must be the finest butterfly ground in the world. The hills are low, and often thickly wooded, with frequent glades deeply shaded, and with enchanting streams cascading over rocks to a little sun-flecked area on the road, where innumerable butterflies congregate to intoxicate themselves on the warm moisture. Such places were met all the way to Bundu Tuan, that is, from 500 to 4900 feet elevation; but if one section is more productive than another, it is the path between Tamu Darat and Kabayau (500 to 800 feet). On the way up I obtained no less than 52 kinds of butterflies on the 13th June, and on our return yet others

10

BUTTERFLIES OF BRITISH NORTH BORNEO. PLATE I.

1, Amnosia decora baluana. M. and F. 2, Lethe darena borneensis. M. and F. 3, Delias eumolpe. M. and F. (see 30). 4, Pareronia valleria lutescens. M. and F. Female mimics 12. 5, Taenaris horsfieldi occulta. M. and F. Male reversed. 6, Synphædra dirtæa. M. and F. 7, Hebomoia glaucippe. M. and F. 8, Discophora necho cheops. M.& F. 9, Danaida crowleyi. Highest altitude butterfly seen. 10, Danaida mellissa microsticta. 11, Danaida luzonensis præmacaristus. 12, Danaida aspasia. 13, Euplœa deione zonata. 14, Euplœa ægyptus. 15, Euplœa diocletianus. M. 16, Euplœa mezares aristotelis. F. 17, Danaida lotis lotis. 18, Hypolymnas misippus. M. 19, Cethosia hypsia hypsia (Dwarf). 20, Cethosis hypsia hypsia. 21, Apatura parisatis borneanana. 22, Huphina hespera. M. 23, Elymnias pellucida. 24, Prothœ frankii borneensis. 25, Papilio antiphates itam puti. 26, Dèrcas verheullii. 27, Venessa canace maniliana. 23, Cyrestis maenalis seminigra. 29, Thaumantis lucipor. M. 30, Delias eumolpe. F. (Reversed) (see 3). 31, Prioneris vollenhovi. Wall. 32, Appias pandione whiteheadi.

11

were found. Amongst these were two beautiful Morphinae—*Zeuxidia doubledayii*, with a blue blaze; and the male *Discophora necho cheops* (Plate I, 8), a brown beauty shot with blue, and with dashes of heliotrope on the fore-wing. *Euripus halitherses borneenses* (Plate II, 18), of which the male is black with clear white splays, occurs; also a small purple butterfly *Mycalesis orseis*, whose brown reverse is margined with lines of 'eyes.' With these were found two handsome moths, a Zygaenid, *Erasmia namouna*, brown with white splays; and a Lithosid *Neochera marmorea*, steel-blue, with the veins exquisitely picked out in white. From arrival in camp till bed-time I was busily employed pinning out the day's captures. It was quite impossible to make more than a casual examination at the time, but the greatest care was taken to arrange the specimen boxes in geographical order for subsequent scrutiny. In this way I was able to prepare the 'table of distribution' given in Appendix V. The table cannot be regarded as complete, but at any rate it affords a fair notion of the range of the various species. The type of Lepidoptera that occurs in these enchanted regions may be seen at a glance by referring to the *Index of Butterflies* and the supplementary list given in *Appendix VI*.

One of the most notable butterflies caught between Tamu Darat and Kabayau was *Taenaris horsfieldi occulta* (Plate I, 5), which is exceedingly rare, and is even worth a good deal of money in England. The colouring is clear dark-grey, turning to fawn on the hind-wing, where, on the upper side, there is one large yellow spot, with a dark, purplish centre; and on the reverse two such spots. The male in the illustration is reversed. I believe a form of this butterfly occurs in the Malay Peninsula, but I have never met it. It is rarely seen outside Museum collections, yet we obtained no less than thirteen specimens, for here it is not at all scarce, and is found all the way to Bundu Tuan. The female is considerably larger than the male.

Here, too, appears the gorgeous peacock-coloured *Papilio, Karna carnatus*, which is one of the glories of Borneo, and which lent a special excitement to the chase all the way to Kaung. All the splendour with which butterflies are endowed seems to be developed in this beauty. It has a span of $5\frac{1}{4}$ inches; the fore-wings, the upper half of the hind-wing, and the superb swallow-tails being black, richly peppered over with a green frost. Across the centre of the hind-wing is a blaze of peacock-blue, with violet and purple reflections according to the light. The reverse is dull black, with whitish splays on the forewing, and a marginal row of violet and magenta 'horse shoes' on the hind-wing, of which one shows through on the upper side. Nothing could possibly be more splendid. The shining green *Papilio palinurus* (Plate II, 12), which also is plentiful, and which had stood hitherto for me as all that is most resplendent, is entirely outclassed. *Papilio karna carnatus* is presented in Plate II, 4; but it will be readily understood how inadequately a photograph illustrates the blaze of rich colour that characterizes many of the butterflies here described. *P. karna carnatus* is represented in Burma by *P. paris*, and by *P. arcturus*, which is not very distantly related.

On this march and the next we fell in with most of the great Morphinae—those large and brilliant denizens of the forest-deep, that are more exciting and more sporting than almost any other butterflies. Between Tamu Darat and Kayabau I caught no less than four kinds, namely, the shining blue male of *Zeuxidia aurelius* (Plate II, 15) (the female, which I have caught in the Peninsula, is brown with bold white splays); the blue-black *Kallima inachus buxtoni* (Plate II, 14), with an orange blaze (invisible in the photograph) across the fore-wing, and a 'leaf-like' reverse; the brown *Thaumantis noureddin* (Plate II, 13), with a superb violet sheen; and the male of *Thauria aliris* (Plate II, 8). Each one of these is a prize in itself; and in such company the capture of a rare and exquisite *Terinos fluminans*, of *Limenitis procris agnata*, *Hypolym as antilope anomala*, and of the large ghostly *Hestia Iynceus fumata* (Plate II, 11), the black and white wraith that floats through the forest, is comparatively tame. I was, however, pleased to get *Ixias pyrene undatus* (Plate II, 17), a sulphur-yellow butterfly with a black tip and an orange blaze. I once caught *Ixias birdi* in the forests of Pahang.

A notable find, even on a day like this, was the female of *Pareronia valleria lutescens* (Plate I, 4). This is the rare black and yellow female of one of the very commonest sea-green butterflies. This female mimics *Danais aspasia* (Plate I, 12), and I had sought the Peninsula form, *P. hippia*, in vain for two years. Between Tamu Darat and Tinompok I obtained three specimens. In the Malay Peninsula the female *P. hippia* escapes attention, no doubt, by the very perfection of its mimicry. But here, in Borneo, the model *(D. aspasia)* is almost scarce, and the mimic consequently more conspicuous.

13

The scarcity of *Euploea* and *Danais* in these parts is really rather extraordinary. The only place where *Euploea* were numerous was in the gorge near Tenom. They are quite the commonest butterflies in the Malay Peninsula. Only three or four species were found, and of *Euploea diocletianus* (Plate I, 15) only one single specimen was seen in the Kinabalu area, and that at Kabayau. The specimen is a male, and is a modification of the Peninsula form, having less white on the fore-wing, and the white splays of the hind-wing much reduced.

Now this suggests something very interesting, for the mimic of *E. diocletianus* on the Peninsula is *Papilio caunus aegialus*, and a female specimen of something close to that which I obtained in Sarawak, and which is identified as the rare *Papilio paradoxa telesicles*, is also modified. Is it that the mimic in Borneo has copied the change in the model? Or is it that the *Paradoxa* here derives little advantage from its mimicry of so scarce a butterfly, and has therefore departed from its mimicry? I am inclined to favour the first alternative, namely, that the Bornean mimic has copied the change in the Bornean model; for in the Sarawak female *Paradoxa* the white on both fore- and hind-wings is greatly reduced, as if to adapt it to just those very changes in the *Euploea*. But with only one female specimen of the Sarawak *Paradoxa* it is impossible to arrive at a definite opinion. It is merely an additional mystery—one of those problems in evolution in which the Lepidoptera are so suggestive, and which add so immensely to the delight and value of their study.

Of the butterflies common in this area, mention should be made of *Symphaedra dirtaea* (Plate I, 6), *Cethosia hypsia hypsia*, which seem to vary greatly in size and detail, *Castalius roxus*, the large orange male *Cynthia erota erotella*, the small magenta *Abisara kausamboides tera*, the small blue-edged *Euthalia cocytina ambalika*, *Curatis malayica*, and sundry 'Blues.' *Euriboea (Eulepis) athamas uraeus* occurs at low levels, but no other species of the family were seen anywhere. *Dercas verheullii* (Plate I, 26), a handsome square-cut yellow with bronze tips, which is comparatively scarce in the Peninsula, occurs in North Borneo in large schools. *Leptocircus curius* (Plate II, 21), whose hind-wing is practically nothing but a long, frail streamer, is seen positively in clouds of fifty and a hundred together. In flight, the white line of the streamer is vibrated, and looks like a string of pearls. It is a beautiful sight to watch flocks of these dainty little *Leptocircus*, which are usually considered rare. *L. meges* (Plate II, 20), which has a blue band across the fore-wing, was seen (but not caught) at Bundu Tuan, and one specimen was secured in the gorge of the Padas River, between Beaufort and Tenom. It does not seem to be common in Borneo; but the profusion of *L. curius* beats anything I ever saw.

Chapter III

THE APPROACH—KABAYAU TO DALLAS

The thinness of the population was increasingly apparent as we advanced. A few Dusun were met going to their markets, or carrying down large baskets of tobacco for Chinese contractors; but the chief evidence that villages exist hidden away on the hills is the fearful destruction of forests. The Dusun employ the same wasteful system of agriculture as do the hill tribes of Burma. That is to say, they destroy areas of jungle in rotation, and plant their rice on the bare hill sides. 'Wet' cultivation occurs in some parts, but there are no terraces in all the country we visited. After a rice crop, the slopes, according to locality, are planted for a season with tobacco or tapioca *(obi kayu)*, and then stand idle till brought into use again after a cycle of about seven years. In this way the hills are denuded of forest over large areas of Dusun country, and the game has been driven away. The Dusun, therefore, have developed into agriculturalists, while the Muruts of the Interior are still essentially hunters.

On the 14th June we marched 11 miles from Kabayau to Kaung through exceedingly pretty country. The hills are rather higher, and in spite of deforestations the jungles are still thick in places. As on the former stages, there are frequent and charming glens with little streams burling down over rocks. We shot a fine black squirrel, a dark form of *Sciurus prevosti (pluto)*, with chestnut belly and paws. It was 19 inches long.

We also got two Broadbills. One, the 'Black-and-Yellow Broadbill' *(Eurylaemus ochromalus kalamantan)*, resembles the Peninsula type, but has perhaps a little less of the beautiful crushed-strawberry colour on the breast. The other, the 'Black-and-Red Broad-bill' *(Cymborhynchus macrorhynchus)*, has the whole upper plumage black, with a broad crimson collar round the throat and running back to below the eye. The breast, belly and rump are also rich red; the tail black; and the inner wing feathers white. The mandibles are turquoise blue with yellow on the

17

under part of the lower one. The eyes are emerald green, and the feet cobalt blue. On the shoulder there is a touch of orange. Some of these Broadbills are very beautiful birds, but when they are stuffed the turquoise beak quickly turns black.

As regards insects, we caught *Tacua speciosa*, a brown Cicada, with the veins of the wing picked out in bronze, and with blue, green and scarlet on the body; and a greenish beetle, $1^1/_2$ inches long, with three blue bands across the wing cases. The green antennae are ornamented at intervals with thick tufts of black hair. By this time news had gone forth that a lunatic had arrived who paid as much as five cents for an insect. The villagers became quite demoralized, and the old men and boys of Kaung brought in all the vermin with which their coconut trees appear to be infested. It is astonishing that any coconuts remain, for amongst handfuls of Weevils was one, *Protocerius colossus*, four inches long, the largest I have ever seen. They also brought me many fine Rhinoceros Beetles *(Trichogomphus milon)*, and a Stag Beetle *(Odontolabis gazella)*, with the body alone $2^1/_2$ inches long. The speculators of Kaung very soon experienced a serious slump in local products, but to my delight produced three *Rhasinids*, those queer, large insects all covered with thorns, about which practically nothing is known. In the Malay Peninsula I have obtained at least one new to science. The three specimens produced here were all in an immature stage, the largest $3^3/_8$ inches long; but one at least resembled a Peninsula type. Besides all these, at least six different kinds of brilliant emerald-green beetles were obtained, of which the most curious was a Scarab, *Theodosia westwoodi* (Plate II, 26) with two enormous horns, one bent forward from the back, and one curling up to meet it from the snout. The 'Stink Bugs' (Pentatomidae) were interesting on account of their great size, their high colours, and their general resemblance to Malay Peninsula types, as in the case of *Eusthenes robustus, Pycanum rubens*, and (Coreidae) *Prionlomia*. Two kinds of the mysterious *Malacoderm*, or *'Trilobite Larvae,'* occur; one brown, like the Malayan form, and one pointed with orange. They were brought in by the villagers literally in handfuls. The curious thing about them is that, though they are exceedingly numerous on Kinabalu, these larvae have never been traced to their adult stage. Shelford, who has kept them alive for months, suggests, in his *Naturalist in Borneo* (page 169), that perhaps the larvae undergo no final metamorphosis at all, but that they are already adult, and breed in this larval form. In support of his theory, Shelford recalls the fact that the female of *Phengodes hieronymi* (a South American *Malacoderm*) has no final metamorphosis, or rather that the adult is indistinguishable from the larva.

Amongst the beetles and weevils brought in by the villagers of Kaung were some larger than any I have seen in the Peninsula, but on the whole, I think a similar expedition there would be more productive of large and curious insects than is Borneo; though whether that is a recommendation from the point of view of ordinary mortals is doubtful! The naturalist, being generally considered slightly mad, regards these things differently. When I proposed catching butterflies in Miri (Sarawak), where they think only in terms of beer and oil, I felt a black suspicion immediately settle upon me. The Sarawak Government formally asked me for a list, not, I fear, from a thirst for knowledge, but imagining, I suppose, that I was hot on the trail of the Gold Bug. Shades of Wallace, Burbidge, Beccari!

Between Kabayau and Kaung the great 'Bird-winged' butterfly *Ornithoptera brookiana* (so named after one of the Brookes of Sarawak) makes its appearance (Plate II, 1). It is velvet-black, with a brilliant splay of green diamonds across the wings, besides touches of deep red, and lines of shining blue. It is the King of Malayan Lepidoptera. While prizes were being showered thick upon me, I hoped in vain for the female *Brookiana*—the most elusive butterfly in the world. Once I have seen her, but she keeps to the tree tops and can usually only be shot down.

We did, however, continue to catch the finest Morphinae, including several varieties of *Thaumantis noureddin*: the shining blue *Thaumantis lucipor* (Plate I, 29), and the small brown *Stibochiona schoenbergi* (Plate II, 19) with fawn marginal plates, each studded with a purple dot. The last is essentially a Kinabalu butterfly, and rare even there. A still more pleasing capture was that of *Prothoe franckii borneensis* (Plate I, 24), which is indigo with a cloudy, cobalt blaze across the fore-wing. Its shape also is most unusual, with the tails curved outwards. The reverse is wonderfully patterned in a design of black, grey and green. A truly gracious butterfly this, with the fleecy clouds and the blue of Heaven on its wings. The crowning success was a lucky 'right and left' at a male

BUTTERFLIES OF BRITISH NORTH BORNEO. PLATE II.
1, Ornithoptera brookiana. M. 2, Ornithoptera amphrysus flavicollis. M. 3, Ornithoptera amphrysus flavicollis. F. 4, Papilio karna carnatus. 5, Papilio memnon sericapus. M. 6, Papilio memnon isaka. F. 7, Papilio helenus. 8, Thauria aliris. M. 9, Thauria aliris. F. 10, Papilio nephelus. 11, Hestia lynceus fumata. 12, Papilio palinurus. 13, Thaumantis noureddin. 14, Kallima inachus buxtoni. 15, Zeuxidia aurelius. M. 16. Pathenos silvia lilacinus. 17, Ixias pyrene undatus. 18, Euripus halitherses bornensis. 19, Stibochiona schoenbergi. 20, Leptocircus meges. 21, Leptocircus curius. 22, Unidentified. 23, Unidentified. 24, Unidentified. 25, Baprestidæ chrysochroa opulenta. 26, Scarabædæ theodosia westwoodi.

and female *Thauria aliris* (Plate II, 8 and 9). At rest these great and glorious butterflies of the deep jungle are invisible, but a number of them suddenly sprang into being in the undergrowth, giving us for some minutes a really exciting chase. This is the sort of trophy of which a hunter may be proud. The Bornean *Thauria aliris* costs as much in hard work, expense and organization as a tiger skin—perhaps more. The female is no less than six inches across the wing—dark-brown shot with blue, and turning to rich orange-yellow on the hind-wing. Across the fore-wing is a very broad white bar. In its way, this magnificent *Thauria* is one of the handsomest of all butterflies, as also it is one of the most difficult to find in its jungle retreat.

The habits of these great, skulking Morphinae, or as they are now, I believe, called, 'Nymphalinae,' have to be very carefully studied. As a rule they spend the day motionless in the depths of the forest, the brilliant wings closed, and showing only the dull, leaflike under-surface. *Kallima inachus buxtoni*, when at rest, is the most perfect imitation of a leaf. A stone thrown into a glade may cause them to betray themselves. They do not commonly occur in primeval forest, or if they do, are unapproachable, but resort rather to secondary jungle where, of course, the undergrowth is even thicker. At regular hours—about 9.00 a.m. and 4.30 p.m.—they sport furtively in the open space of the path for a few minutes. And again at dusk in certain favourable spots I have watched them against the sky chasing each other at high speed. But it is then too dark to see them in the foliage.

It is a red-letter day that introduces one for the first time to some of the charming butterflies I have here described. I think I shall never forget those brilliant mornings at the base of Kinabalu when each bend in the path, each mountain stream, each vista of jungle, might reveal, and generally did reveal, some new and unexpected wonder. The circumstances made it necessary to catch as many as possible in a limited time, since it is often desirable, for purposes of study, to obtain a 'series' of this species or that. But in me there is no lust for capture after sufficient specimens have been secured for my purpose. It pleases me enough, once I know them, merely to see rare and beautiful creatures moving fearlessly in their natural surroundings. Many must deprecate, as I do, the attitude—"A fine day: let's kill something"—and it may interest those not too scientifically-minded to note my method of collecting.

Each new specimen has a page, and is described and numbered in a 'Field Note-Book,' and is later painted under a corresponding number in a 'Sketch-Book.' This is a laborious process, but very instructive; and it creates a permanent, portable record that is easily referred to, while the actual specimens, alas! fade with time, and are subject to sad vicissitudes from travel, ants and damp. Already those gorgeous wings from Borneo have lost something of their fresh beauty; but in my Sketch-Books they live and restore to my eyes a dreamland of romance that belongs to memories of Kinabalu.

At Kaung, negotiations began regarding porters and guides for the ascent of Kinabalu. This is a protracted business, an inevitable phase of all expeditions, for in the Dusun we are dealing with very primitive people to whom it is necessary to disclose a plan bit by bit. They cannot grasp the whole matter all at once. Suppose you want rice, as we did now, and you said:—

"Have you got rice?" the reply would be "No." But since they are known to have rice, the subject should be approached deviously, thus:—
"I see you have cleared new ground this year."
"I suppose men come back from the estates to work?"
"Had you a good crop?"
"Did you get plenty of rice?" (Here the unsuspecting Dusun commits himself.) "We want rice."
"We want twenty *kattis of rice.*"

The result is that rice is supplied. One has to remember that some of these Dusun have never been further from their homes than Kotabalud. There is no world for them beyond that.

Yet, small though the population is, the existing communities, because of their wasteful agricultural methods, have had to split up from time to time. These colonies found new villages, calling them after the old ones. This may account for the duplication of names and also for the dwindling of villages in size. Kaung, which is now quite a small place, is described by Low in 1851 as "certainly containing not less than two to three hundred houses."

Kaung, being a handy village, I wanted to see a 'head-house.' This is rather a delicate subject but Round-Turner is an expert at going sideways at a thing. He began:—

"Are you well?"

"No sickness here?"

"Lived here long?"

"Were you here during the Wars?"

"Were you brave enough to get any heads?"

"Where are the heads now?"

It seems that the heads are not kept at Kaung. Either they were removed by one of our punitive expeditions, or they were bought by the Dusun of Bundu Tuan who still like to have a nice pile of human skulls in the middle of the floor when they have a drunk. But Whitehead says, writing in 1887:—

"The villagers of Kiau and Koung, and surrounding districts, are decided head-hunters and I believe make annual expeditions."

The Dusun suffer a good deal from skin diseases, which are common amongst wild and ill-fed folk. In the old days of tribal wars a good itch that kept one wakeful was considered an asset: and, it is said, a man with a really efficient scratchiness could sell the infection for a high price— say a chicken.

On the whole they are good people, these Dusun, with considerable sense of humour—the sort of folk one could do a lot with by cultivating them. Few Europeans seem to have learnt their language, which is, of course, the first step towards winning their confidence. They suffer, too, from a multitude of dialects, the result of isolation arising from former head-hunting customs. They are still very ignorant, and have not reached the stage of wanting to learn; and the school started for them in Kotabalud has been closed. But they enter the Company's police service freely, which is, after all, the best possible school for a people in their situation.

As an ex-recruiting officer, I was constantly struck with the resemblance of these Dusun to the various hill tribes of Burma—Kachin, Muru and Chin—who are every bit as backward, and whose employment in the regular Indian Army has proved such a remarkable success. More Dusun present themselves for the Company's service than can be accepted, for the local forces consist of only about 800 men, of whom half are Indians. The thread-bare cult of the Sikh, now exploded even in the Federated Malay States, lingers on here—with all that it entails of corruption and money-lending. Borneo at various times has sought soldiers amongst Somalis, even amongst Indian exmutineers! neglecting the excellent military material available in the country whose employment would do more than anything else to cultivate and civilize the backward races. As a Civil Officer in Burma said to me on the return of my Kachin Company from Mesopotamia:—

"It has been worth more than fifty years of administration to these Hills."

And Borneo possesses in its Dusun, Murut and Dayak, the same simple, sturdy, manageable, Mongolian type that is largely recruited in Burma now, and which is, after all, first cousin to the Gurkha himself.

On the 15th June we marched 9 miles to Dallas, the bridle-path now rising into higher hills with a wonderfully easy gradient. Dallas is about 2500 feet, and the air is cool and refreshing. Rain storms may be expected every afternoon at this season, but as a matter of fact we were entering upon a patch of bad weather, and there was a good deal too much rain after mid-day. The mornings, however, were brilliant. The hillsides round Dallas are almost bare, so extensive has been the damage to the forest caused by 'Ladang' cultivation. The steep, treeless slopes, with little watchhouses in the clearings, recalled similar conditions amongst the Kachins and Yawyins of Burma. Generally speaking, the vegetation is Malayan. Bamboo had not yet begun to occur, and flowers of all sorts, even orchids, were scarce; but bracken *(Pteris arachnoidae)* makes its appearance, and on the hillsides grow giant ferns with fronds 15 feet long.

At Dallas the rest-house is boldly situated on a bare spur of mountain. Before it the valley falls away in an abyss, and beyond, Kinabalu, now close at hand, rises in all its savage grandeur. This aspect of it is magnificent. The peak is vast, naked, threatening, lonely; its stupendous precipices are laced with cataracts—not less than fifteen of them, some of which must be almost a thousand feet high—white ribbons of foam against the dark granite. After rain the mountain spouts water. The gleam of its wet surface has been taken for snow, though, in fact, snow is not known to lie. But as Sau Nan, my Kachin orderly, said as we gazed up towards Kinabalu from Dallas:—

"It is not the height that is so astonishing: but the nakedness of the rock."

To picture the scene you must imagine a foreground of long, grassy ridges, gently rising and beautifully green. The village of Kyau lies on those slopes, and a bridle-path to it can be seen. Above these ridges rise the lower spurs of Kinabalu itself, ascending in cones of densely-wooded upland, laid like buttresses against the mountain. And from these again rise smooth, menacing walls of granite, ending in the curious pinnacles, bastions and towers that are so characteristic of every aspect of Kinabalu. It is truly an impressive picture that cannot fail to inspire respect and wonder in one about to ascend to those apparently inaccessible heights.

In the sunset Kinabalu sometimes assumes the most beautiful colouring. The "alpine-glows" as seen from Dallas are often brilliant: but this evening the mountain assumed a dark, forbidding mood that refused to be illuminated, though away in the west, where a little patch of sea is visible, the sun sank in a glory of gold and carmine clouds. At least, I think it was in the west; but in Borneo my sense of direction, usually good, was completely bewildered. This may be due to the fact that Kinabalu lies aslant north-west and south-east, or to the curious configuration of the coast. But certainly at Benoni the sun sets habitually in the north, into the middle of the China Sea.

We halted at Dallas on the 16th to conclude arrangements for twenty porters, two guides and a priest, who were all recruited from the opposite village of Kyau. They were a sturdy, sensible lot of Dusun, who, as it proved, worked exceedingly well. It was agreed that each porter should receive $3, a gangtang of rice, and a blanket. (I had brought a load of blankets on purpose.) The guides, Lamat and Surea, were to get $10 each; and the priest, whose ministrations are considered absolutely necessary, $6 and a porter for his kit. The priest was a good old fellow. I cultivated his good offices with whisky, and in return he kept the Spirits of the Mountain content. The nature of the sacrifices and offerings as again stated—seven eggs, seven chickens, a salute of two shots at the foot of the last climb, and two more at the summit, though by now, after

25

a running haggle lasting three days, I was not likely to forget it. It is a tradition with Kinabalu expeditions to fight the rapacity of the villagers, which indeed is only kept within bounds by an 'offensive-defensive.' The time spent in battling over 77 cent's worth of sacrifice might, if less skilfully used, be directed to something far more expensive—such as Norfolk jackets for the guides, or a pair of trousers for the priest at the very least. Alas! the days are gone when they might be satisfied with an empty jam tin.

On the whole, the middle of June is a favourable season for an ascent of Kinabalu. But the last week in May would be better, for the villagers would still be at leisure, and the weather more likely to be settled.

Chapter IV

THE ASCENT

Round-Turner, having made all possible arrangements, returned to Kotabalud, and I launched out alone for the ascent of Kinabalu, following the bridle-path for another nine miles to where it bifurcates— one branch going to Renau, and the main one to Bundu Tuan, and so, after about another ten days' journey, to Tambunan, Keningau and Tenom in the interior. Just at this bifurcation is the low pass of Tinompok (4900 feet), across the ridge by which the ascent of Kinabalu is made. Here we intended to build a shelter and sleep; but the Dusun pointed out a couple of huts half a mile away toward Bundu Tuan, where we spent the night very comfortably.

After leaving the bare hills of Dallas, we had reentered big forests that extend to Tinompok. The elevation is about 4500 feet, and the character of the jungle undergoes a marked change. Bamboo is more plentiful; tree-ferns make their appearance; and the great trees have a tendency to be buttressed at the base. The rain-drenched hills are rich in ferns and mosses, of which several specimens were collected. Flowers, which had been rare hitherto, began to appear. Near Dallas we got a large head of bloom like that of a foxglove; and in the Tinompok forests two orchids— one with a little white flower, the other with a fine cluster of four scarlet trumpets.

We were now definitely on the base of Mount Kinabalu, and the butterflies immediately became interesting. The appearance of the white, black-edged *Appais pandione whiteheadi* was very abrupt, these butterflies flying in little schools, but quickly, so that they were rather hard to catch. The transition from the normal *Papilio antiphates itamputi* to *P. stratiotes*, and from *Ragadia crisia siponta* to the specialized Kinabalu form, *R. melindena annulata*, has already been referred to. Other species that make their first appearance are the little brown *Apatura parisatis borneanana* (Plate I, 21), the black, orange-striped

27

ONE OF THE MANY PEAKS OF MOUNT KINABALU ONLY A LITTLE LOWER THAN THE MAIN PEAK.

Symbrenthia hypselis balunda, and the indigo *Venessa canace maniliana*, which is crossed with cobalt-blue, and has a very curious outline (Plate I, 27). The last two are beautifully variegated on the reverse. The handsome Bornean form of *Papilio helenus* (Plate II, 7) occurs; and also *Liminitis daraxa viridicans*, which is blackish-brown with a line of small green plates right down the middle of both wings.

We worked through these forests very carefully on several occasions, and came to the conclusion that the Kinabalu forms appear quite suddenly somewhere between Dallas and the Tinompok Pass, the destruction of a belt of forest having now made the dividing line very clear—namely, at a point about three miles beyond Dallas where the jungle begins again. It is interesting to speculate as to the reason for the very limited distribution of the Kinabalu forms. It cannot, of course, be the gap in the forest that now *seems* to constitute the boundary, since the vegetation can only have been cleared in comparatively recent times. It cannot be due to elevation, since the butterflies of Kinabalu, as we shall see later, are nearly all found at its base up to 5000 feet, and there are many other uplands in Borneo higher than that. The specialization of the types, then, can only be accounted for by some marked period of isolation when most of Borneo sank below the waves of the sea, leaving such high regions as Kinabalu the only sanctuaries for the ancestors of the peculiar plants, birds and butterflies that inhabit them today. This supposition seems to be strengthened by the recent observations of Dr Mjöberg, that some at least of the types hitherto supposed to have been peculiar to Kinabalu occur also along the central ranges as far as the peaks of Murud and Tibang.

Many and beautiful are these types. We can only mention a few here, such as *Elymnias pellucida* (Plate I, 23), a dark-brown butterfly with touches of white arranged in little etched triangles upon the hind-wing; and *Tanaecia amisa*, which is blackish-brown with a row of white plates, making a broad, continuous line down the middle of both wings. Both are peculiar to Kinabalu, and extremely rare even there. *Lethe darena borneensis* (Plate I, 2) is a rich, warm brown, turning to chestnut on the hind-wing, though these fine colours are not well represented in the illustration. The rare female is chestnut, with a yellow blaze and black tips. These are the most beautiful of all the *Lethe*, and they are also an extreme case of dimorphism, or dissimilarity between the male and female. An exceedingly beautiful rose-coloured Zemeros—*Laxita lola*, occurs. The dark brown *Amnosia decora balauna* (Plate I, 1), with a bright cobalt blaze, is plentiful, and it has a wider range, being found as far down as Kabayau. The very rare female, of which only one specimen was obtained, is brown with a white line across the fore-wing, and a marginal row of 'eyes' on the hind-wing. Here, again, no one would suspect any relationship between the male and female. A peculiar Danais, *Mellissa m icrosticta* (Plate I, 10) is almost black, with thin, pale blue splays; and yet another, *Danaida crowleyi* (Plate I, 9) was found at a greater altitude (7000 feet) than any other butterfly.

As for the insects, the porters on arrival at Tinompok set themselves to earn five-cent pieces, and positively inundated me. Here, and later at Bundu Tuan, a very large collection of spiders was made, till every test-tube was full of the nasty things. Nevertheless, many are really quite beautiful—some being scarlet, some blue and scarlet, black and silver, and silver and green *(Leucauge)*. The little 'Crab-Spiders' are of many colours—brown, chestnut and yellow. The highest-living insect taken was an *Opilione* at Paka Cave (9790 feet). At the time of writing, the most interesting specimens are a fairly large burrowing spider *(Coremiocnemis)*, which appears to be new; and a probably new species of the family Argiopidae, of which the genus *Clitaetra* has hitherto been reported only from Madagascar and Ceylon. But the collection, now in expert hands, will probably take months to work out. The first step is to reduce the individuals to their genus, after which the species can only be determined by reference to a scattered literature in various languages. Spiders are not creatures of which I know anything, but a report on them by Mr Abraham will be found in Appendix III.

Near Tinompok, Mingga, the Dayak collector, shot a tiny greyish squirrel, *Glyphotes simus*, striped darkly on the flanks. The Dusun call it *'Montuk.'* Including the tail it is only eight inches long. This is the maximum size, and it is peculiar to Kinabalu. We saw nothing of the other pigmy squirrel, *Sciurus whiteheadi*, which Whitehead describes as "being brownish-yellow, about the size of a large mouse, and with the ears ornamented with long tufts of grey fur." Very common here is the curious Rain Bird (evil omen), *Dendrocitta occipitalis cinerascens*, that floats through the jungle like a herald of wet weather. The upper plumage is greyish-blue, the lower dusky-brown, turning almost chestnut on the belly. The wings are blue-black with white spots on the primaries. The tail is black, but covered at the base by blue-grey upper covers, and by long, thinly-webbed quills ending in two black extensions that stick out behind the bird, as do the quills of the Burmese Racquettailed Drongo.

The huts in which we were housed below Tinompok were beautifully situated on a grassy slope. On one hand rose the virgin forests of Kinabalu, and on the other the ground fell away to the village of Bundu Tuan, which could be seen far below. In front, across a valley, spread a lovely panorama of quiet, open hills.

Kinabalu now lay end on, and looked considerably easier. Its flanks rise in the usual precipices, but the ridges we were to follow seemed to offer a fairly reasonable line of ascent. Still, the reputed 'difficult bit' at the bottom of the final slope made me a little apprehensive, because now it seemed as if the climb was the real object of the Expedition, and that success in that would set a mark upon all our labours. Moreover, the Government was so obviously anxious for the Expedition to reach the summit, and had taken so much trouble to render that possible.

The actual ascent takes three days. On the 18th of June the mountain was cloudless and we set out, plunging into dense jungle, and climbing in three hours to camp at Lumu Lumu. The path, which seems to be used by the Dusun for bringing down jungle produce, is rough, but pretty distinct. Indeed, you cannot leave it for the thickness of the undergrowth on either side. It clings to the crest of the ridge all the way. You can picture the rises mapped out from below as you climb stiffly and then ease off for a little; but in this smother of vegetation nothing whatever can be seen. No gay butterflies invade these gloomy solitudes, which are given over to the bloodthirsty leech. Few birds are met with, but we were lucky in procuring a female of the peculiar Kinabalu Trogon— *Harpactes whiteheadi*, or, as it is now called, *Pyrotrogon whiteheadi*. The beak and cheeks are bluish, the back is chestnut, and the lower back closely barred with black and brown. The wings are black, with the outer edge of each feather finely lined with white, and with the base of the inner web also white. In some of the tailfeathers the ends are black, in others white, and in yet others chestnut. The breast is grey and the belly chestnut. The male, which we did not see, is altogether more splendid.

The only other objects of interest were the mosses, of which I found one new species of *Dicranoloma angustifrondum* Dixon, and one new variety of the species *Hypnodendron copelandii* Broth. *(latifolium)*. The latter species has hitherto only been known in the Philippines. In these rain-drenched jungles the ground, the trees, and even some of the leaves, are covered with moss. I collected no less than eighteen distinct kinds— some very pretty and quaint, especially one which grows like a mushroom on a long stalk. All these I placed in sodden blotting-paper, and hoped for the best.

We reached Lumu Lumu at 10.00 a.m. It seemed absurd to go such a little way, but the necessity for short stages was soon apparent. To begin with, the porters were laden, and they are not great weight-carriers; shelter had to be provided; and rain may always be expected in the afternoon. As it was, we were not settled down till tea-time. The huts used by the last Expedition, that of Evans and Sarel, in April 1924, had collapsed, and we had to build fresh ones. There is no foliage suitable for quick roofing, and the bark of a tree has to be stripped and laid on. It takes time. Then the ground was saturated, and the fallen timber rotten and soggy, so that it was only Sau Nan's jungle lore that enabled us to kindle a fire. And even before the huts were roofed down came the rain that, as the Rain Bird had foretold, was destined to last for thirty-six hours. It was now that I had cause to congratulate myself on having brought a couple of Kachins from the forest hills of Burma. In the meantime birds and botanical specimens had to be cared for. Quinine and whisky were issued, and we settled down to a thoroughly wretched night. The discomfort of these camps may be gathered from the fact that it was four days before I got a bath. The altitude of Lumu Lumu is about 5500 feet.

Next morning it was still raining, and the porters demurred at marching, saying that in similar conditions two had died on a former occasion. It was the only time I had to be stern to them, and eventually we got started in fair time, considering the state of the camp. The climb of five and a half hours to Paka Cave is a hard one, and it was only during the descent that we realized how very steep the path is. In places the ridge, to the crest of which the path clings, is very narrow, with steep slopes on either side; and the presence of vegetation veils the aspect of many little bits that might otherwise appear unpleasant. As it is, the only trouble is with

32

regard to altitude, which, after such violent exertion, began to tell, as the party climbed rapidly to nearly 10,000 feet. Sau Nan was unable to sleep at this altitude. I take my hat off to Miss Gibbs, the first lady to make this ascent; but the journey is not to be recommended generally for women.

This day's climb may be divided into two phases—the first up to Kamburonga (about 7040 feet), the second to Paka Cave. The first was through the same sort of dense, moss-grown jungle as on the preceding day, but with steeper slopes, and never an instant's respite. It was a matter of stepping along roots and fallen tree trunks. Above Lumu Lumu there is a zone of flowers, mostly orchids. It is of short duration, but the first salmon pink rhododendron seen there extend far up the mountain till they are replaced by a red variety. Later come horse-chestnut. On one tree we saw cut the names of Evans, Sarel and their guide Umpoh, and the date 20.1.24. I suppose they carved the record on the first suitable wood on their way home. At Kamburonga we stampeded pig, and from the tracks it would appear there were many. Evans and Sarel in their report speak of seeing the trail of a rhino at 8000 feet.

Kamburonga is a Dusun word which some say means 'Abode of Spirits', but according to I.H.M. Evans, 'Kamburonga' is a mystic plant used for divination, and sometimes hung outside houses to avert sickness. On this wind-swept ridge of Kamburonga, on a swampy space ten yards square, St John camped in 1858, and Whitehead spent six weeks in 1887 making that remarkable collection of birds that established his reputation as an ornithologist. It is here that he caught his highest altitude butterfly, and that I also caught mine 38 years later (Plate I, 9). It is here, too, that an abrupt change occurs in the vegetation, Myrtaceae *(Leptospermum recurvum)* and large *Nepenthes* (pitcher-plants) being met for the first time. Early writers speak of Casuarina being a conspicuous tree, but there are certainly none there now. The first small, green pitcher-plants were obtained at about 4500 feet; but those of Kamburonga are *Nepenthes lowii,* one of the giant forms peculiar to Kinabalu. I can only describe *N. lowii* very inadequately, as like a twisted flagon. The top half, or cup, is oval, and deep brown inside. The lower half, which is twisted to one side, is the pitcher, or receptacle. The 'lid' is covered inside with short thick bristles, and has a white secretion on the hairy surface. The whole plant is green outside, and the cup deep ruddy-brown. The texture is leathery, but at the same time brittle and apt to crack. *Nepenthes lowii* are not found higher up the mountain. As a rule, they rest among the mosses on the ground. The total length in a straight line from the base of the pitcher to the tip of the ' lid ' is 12 inches. If the tape follows the twist, it is 15 inches. The oval cup is seven inches long and $3^1/_4$ inches across. St John considers *Lowii* the most beautiful of all the pitcher-plants and suggests that its twisted shape might be made the model for an elegant claret jug.

The other large kind seen, *Nepenthes villosa*, is altogether more highly specialized. The colour is green and pink, often wholly pink, and the shape that of an egg of which the top has been cut off. On one side of the 'egg' is a column to which is attached the flat, heart-shaped 'lid' in an almost horizontal position. But the most wonderful part is the edge of the 'egg-shell' which is closely armed as with the flanges of a radiator— each flange ending in a sharp spike pointing down and inwards, and forming a *cheveux-de-frise* that prevents the escape of any insect that has fallen into the cup. It must be remembered that these plants secrete a fluid in the pitcher by means of which they digest the animals captured. They are truly carnivorous. In this case, the downward pointed thorns surround the entire cup, and extend also up the column that supports the 'lid.' The inside of the pitcher is creamy-yellow round the top, but deep magenta at the bottom. The 'egg' is $9^1/_2$ inches round. If the lid be treated as quite horizontal, the height of the plant is 71 inches. This species seems to begin at about 7500 feet; and no large *Nepenthes* were noticed above 8500 feet, though our scrutiny, on account of the weather and cold, was necessarily superficial. Whitehead says they extend up to 10,000 feet, but he possibly over-estimated the height, as he makes Paka Cave 10,300 feet. The elevation of Paka, now generally accepted, is 9790 feet. Burbidge notices the first *Nepenthes lowii* at 5000 feet; further up *N. rajah*, then *N. edwardsiana*, and highest of all *N. villosa*. But there is some difficulty in identifying his route.

Four large *Nepenthes* are found on Mt. Kinabalu, of which Kamburonga is rich in these two, *N. lowii* and *N. villosa*. The other two, *N. rajah* and *N. edwardsiana*, do not seem to occur on this part of the mountain. The best place for seeing them, and others, is the Marei Parei plateau above Kyau, which was visited by Low. and St John together. *N. rajah*, the finest of the *Nepenthes*, is named after Raja Brooke of Sarawak, and occurs at 5000 feet. It is the largest of them all, having egg shaped pitchers as much as 19 inches in circumference, in one of which Low found a drowned rat. Hooker describes *N. rajah* as "one of the most striking vegetable productions hitherto discovered, and worthy of taking place beside the *Rafflesia arnoldii*." The cups rest on the ground in a circle, and when old are purplish in colour. *N. edwardsiana* which grows from 6000 to 8000 feet, is said to resemble *Villosa*, having the same fringe of spikes along its lip, but it is more elongated.

It will be noticed that these specialized Kinabalu pitcher-plants do not grow at extraordinarily high elevations. As with the specialized butterflies, which nearly all occur below 5000 feet, there is nothing in their preference for altitude to have localized them so persistently to this one mountain.

The second phase of the climb from Kamburonga to Paka Cave starts, as I have said, with the first of the *Leptospermum recurvum*, whose bent and tortured limbs are the predominant note of the vegetation, and which correspond to the juniper that might be expected at such altitudes in the Himalayas (see illustration on page 82). To begin with, they are fairly large trees, but gradually dwindle in size as the altitude increases, till they are mere shrubs interspersed with red rhododendron and small pine, that Whitehead calls *Dacrydium elatum* (Wall). In fact, the great jungles have been left behind, being replaced by low scrub, and instead of treading on roots, we were clambering up steep ridges of rock. On the way down, when the weather was fine, the views from this section were superb. Above rose the great naked granite summit of the mountain, while far below spread hills and sea, with the islands of Jesselton in the distance. But on our way up all this was swathed in fog. Frequently it rained hard with a chill wind, and the trickling of water down one's limbs was horribly cold. The settled and hopeless appearance of the sky made me very anxious for the success of the Expedition; for not only would it be utterly useless to reach the summit in such conditions, but to attempt it would be dangerous.

I think we were all glad to see Paka Cave. My Kachins, and also the Dusun porters, did their best to keep cheery; but who can pretend to enjoy pouring rain on a bleak mountain-side at 9790 feet? The Cave, which is in fact merely a vast over-hanging boulder, is capable of holding only five or six persons with comfort. Yet all the 23 porters and guides

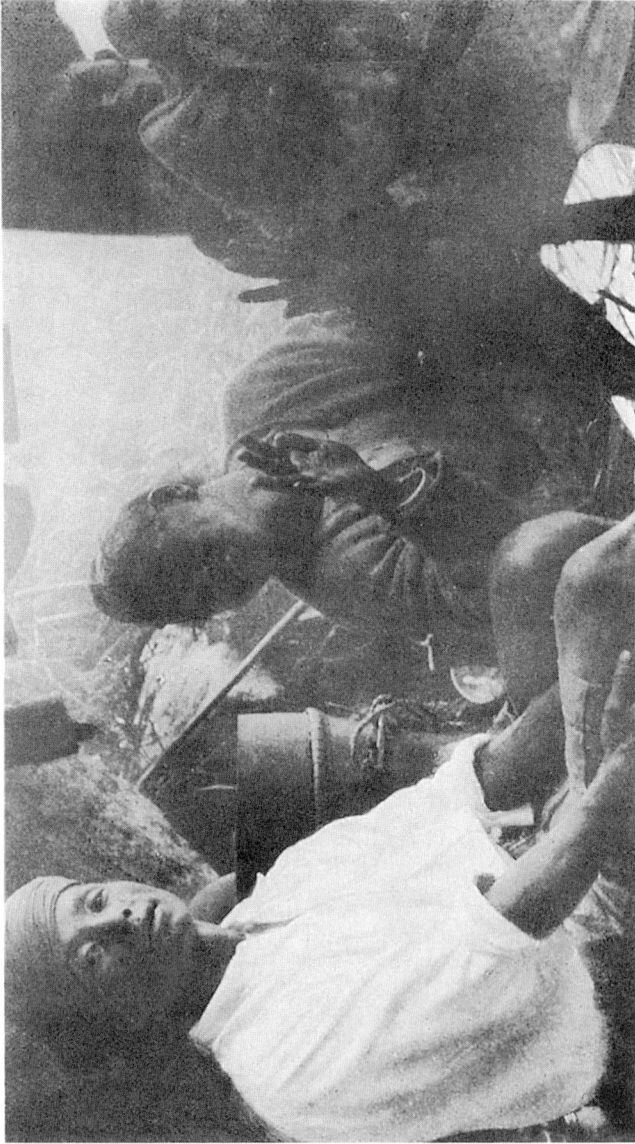

IN PAKA CAVE. *(Sau Nan on the left.)*

36

crowded in too, shuddering with cold, till there was no room to move about or even to sit down. And, indeed, wet as we were, it was bitter. However, we quickly had a fire going. Tea for all was soon preparing, and before it was ready a ration of whisky was issued that vastly improved the aspect of things. Future Expeditions should certainly take up rum or native spirit, since the demands of 25 frozen porters on whisky is devastating. Vitality being restored, a hut for the porters was run up on the path (the only moderately flat place); clothes were hung up to dry, and gradually the teeth of the Kinabalu Expedition ceased to chatter.

Outside the cave there are plenty of trees, and a brawling stream (the source of the Tempassuk River) that on our arrival had changed into a raging torrent. Though there is no flat ground outside, the interior of the cave at least was moderately warm and dry. More comfortable still are the feelings it inspires. For the naturalist and the traveller, this tiny cave at Paka, high up on the mountain side, is classic ground—a goal so elusive, so remote, that it seems like a dream to be there. As I lay at night looking up in the fire-light at the huge boulder that forms the roof, I thought of the select little band, Low, St. John, Whitehead and the rest, who have used this cave during the last three-quarters of a century, and who have left those romantic books of travel and natural history which had drawn me also towards the mysteries of Kinabalu.

Outside, it was still raining in torrents. The last thing I remember before dropping off to sleep was the kneeling figure of Sau Nan praying for fair weather.

BARE GRANITE WALLS RISING ABOVE TREE-LIMIT AT AN ANGLE
OF 40 DEGREES.

Chapter V

THE SUMMIT

On the 20th June I was the first awake; and in the mouth of the cave I saw two little stars. The sky was absolutely clear, and to my very great relief the rain had ceased just in time for our purpose. Nevertheless I carried up stores and blankets, lest we should be detained on the mountain—a precaution that proved unnecessary. Owing to the restricted conditions of the camp, it was 6.30 a.m. before we got clear, but we were on the summit at 9 a.m., after two-and-a-half hours' climb.

The tree limit proved to be hardly 1000 feet above Paka Cave. This spot is now known to the Dusun as Sayat Sayat, though the name seems to have been invented by the Learmonth Expedition for a much higher place where they camped. From Sayat Sayat we looked up to naked granite that shelved up for some 3000 feet, at an angle of 40 degrees. The aspect of these cliffs was almost terribly beautiful in the early morning light; and here at the feet of the rock walls we fired a salute of two guns according to contract. The old priest, who carried a bundle of charms that seemed to consist chiefly of pigs' tushes, scattered rice on the ground, offered an egg and a fowl, and addressed the Spirits of the Mountain. These simple rites at the foot of the great cliffs were not a little impressive. The poor Dusun know not what they fear; yet down in the hearts of the most cynical Twentieth Century being also there lurks that same dread of the overwhelming forces of Nature.

As for the spirits of the Dusun dead—well, let us humour them, like the old lady who bowed at the Devil's name in Church because "It costs nothing to be polite, and, besides, you never know."

Possibly there is something of that attitude amongst the Dusun themselves. At any rate, it is their belief that the souls of the dead tend to linger about their former homes, and to discourage this the bier, and also the steps leading to the house, are sometimes cut up, and the spirits urgently invited to depart to Kinabalu. The routes by which these ghostly journeys are made, says I.H.N. Evans, are well known, and "the passage of the Spirit may be heard at a large stone in the stream at Koraput. The passing of an old man is known from the sound of the tapping of his stick, of a youth or maiden by the music of their guitar, and of a child by the sound of weeping." The Animism of the Dusun, as I have explained, is of a very hazy nature. "To what Gods," I asked, "are these sacrifices offered? "And they replied: "To the *Hantu*—to the ghosts, to the Spirits of the Dead."

The fulfilment of sacrifices is an essential condition under which guides and porters to the mountain are engaged. These rites are particularly specified in the contract—seven eggs and seven fowls (originally they had to be white fowls, but the custom is now relaxed), and a salute of four gun-shots—two at the commencement of the final ascent and two at the summit. No neglect of these formalities is permitted, as we now learned. The twelve-bore cartridges having got wet, we had to borrow the collector's gun. Mingga, who remained in camp, apparently resented this, and supplied only the four cartridges necessary, of which two were fired by mistake at Paka. Two more being demanded at Sayat Sayat, I objected that no rounds would be left for the summit. The salute fired at the wrong place was not, however, accepted as a substitute, and a man had to return all the way to the Cave for more ammunition. For, if these ceremonies were omitted, those ascending the heights would never find their way back again. The ascent is hedged about with various tabus. The names of streams may not be mentioned, and the words 'Kinabalu' or 'Nabalu' may not be used, the mountain being only referred to as 'Agayoh Ngaran,' or 'Big Name.'

Tabus of this kind are common among primitive people. Malays will not use the word 'tiger' in a jungle, and there is a host of restrictions in connection with the collection of camphor. To this day there are forbidden words among Scottish fishermen while they are at sea. Yet these 'superstitions' seem to arise, not from an ignorance of nature, but rather from a close understanding of her more terrible moods. They denote a sense of humiliation in the presence of great forces where perhaps more arrogant people betray a profounder ignorance by ridiculing them. Is it not Montaigne who says:—
"I may well be excused if I rather accept an odd number than an even;

Thursday in respect of Friday; if I had rather make a twelfth or fourteenth at a table, than a thirteenth . . . All such fond conceits, now in credit about us, deserve at least to be listened unto."

The Dusun themselves never go to the summit of Kinabalu of their own accord, partly because of the cold, partly because of the discomforts of the climb, but more especially because the spirits of their dead reside on those frozen heights, brooding over the green valleys and homely villages far below. Nevertheless, these people like to avail themselves of a safe opportunity for going up, and when an Expedition such as this is organized, there are nearly always volunteers among the porters or men who attach themselves to the party.

After the religious formalities at Sayat Sayat we set out immediately for the only troublesome part of the climb. From descriptions of this place I imagined the difficulty to lie in crossing a slope, easy in itself, but with an ugly drop below. It is not that at all, but an ascent for about three-quarters of a mile up narrow slopes of granite just so steep that if you slipped there might follow a long slide with possibly fatal consequences. However, the slopes are not so severe as to make one giddy; nor are they so uniformly smooth that there are not cracks, channels, shrubs and other unevennesses here and there. The ascent, at this point, is certainly a little trying, and the descent even more so; but there is really no danger, except that of growing careless. It was due to carelessness that I here tore off one of my finger-nails. Evans and Sarel in their report recommend rubber soles. Personally, I find rubber highly unsuitable on naked rock. I wore hob-nailed Service ammunition boots, and found they gripped the smooth surface very well. The guides and porters with bare feet had no trouble at all. Much, however, depends on the weather. The rock is naturally slippery when wet, and runs with water. St. John, who came down in rain, says:—

"The descent was a work of danger, as streams crossed our path in every direction; and had we lost our footing while passing them, we should have been sent gliding down to the precipices. It was bitterly cold, the thermometer at 2.00 p.m. falling to 43 degrees; and as we approached the steeper incline the velocity of the running water increased, and the granite became slippery as glass."

THE GENTLY SLOPING GRANITE PLATEAU OF THE SUMMIT.

42

This little section of the climb, in fact, adds a spice of excitement, even of anxiety, but, at least in fine weather, it is one that an inexperienced mountaineer can easily negotiate. It leads up to a sort of plateau, and when that is reached it is all plain sailing.

This 'plateau' is really an extensive area of gently sloping and absolutely naked granite. Large, flat slabs are scattered loose upon it. In one place I noticed a narrow, whitish vein running across the surface of the granite, quite straight, for a long distance. In a few sheltered situations along the edge of the 'plateau' grow little clusters of stunted *Leptospermum*; and in the chinks of the rock, rare tufts of grass and flowers, and a form of dwarf *Leptospermum recurvum* that is only a few inches high; but except for these it is all bare—a gently sloping floor many acres in extent, from the edges of which rise five or six savage looking peaks. Fortunately, the one which is credited with being the highest (though only by five feet) is easy of ascent.

But the summit of Kinabalu is by no means as flat as one might suppose from certain distant aspects of it, and this is soon evident, for on crossing the plateau it is found to end abruptly in a horrid abyss—a sort of 'crater'; a veritable 'Devil's Cauldron' of incredible depth, whose walls rise in sheer precipices for thousands of feet. As the outer aspect of Kinabalu is of perpendicular lines, so also is its interior. This gulf in the centre of the mountain is certainly a grand and terrifying sight.

The actual summit, a loosely piled mass of rock, is on the lip of this abyss. We reached it at 9.30 a.m., and gave a cheer for Burma, which is really our home-land. And since Maung Ba Kye, the Burman, had found the ascent most trying, and had therefore exerted the most pluck in accomplishing it, we gave the bottle that contained the record of our Expedition to him to deposit with those of previous climbers. As I have already said, we found two records that seem to be those of Chinese. There were only four bottles, and it is to be feared the porters steal them. The records of several climbs are stuffed into one bottle that is cracked and without a cork. The next party should take up a wide-mouthed jar to preserve these documents. The custom of leaving a record on Kinabalu seems to have been instituted by Low.

SUMMIT OF MOUNT KINABALU, 13,455 FEET.

44

The summit of loose piled rock on which we stood is called 'Low's Peak' (I3,455 feet). Of the neighbouring heights that rose close round us from the edges of the 'plateau', that towards Kotabalud is Victoria Peak (I3,450 feet); and that towards Bundu Tuan, St. John's Peak (13,440 feet). All these are grouped round the plateau. But further along towards King George Peak (13,345 feet) and King Edward Peak (13,405 feet) the mountain is ridge-like, and apparently unapproachable. The crest of Kinabalu may therefore be regarded as partly plateau and partly ridge, with numbers of summits all of about the same elevation, some of which are mere pinnacles most fantastically shaped.

The point first reached by Low must have been between two of these pinnacles. But the highest summit, now known as Low's Peak, is almost the only one that can be ascended without ropes. St. John seems to have arrived within a stone's throw of the top of Alexandra Peak. Writing of St. John's Peak, that now bears his name, he says:—

"From the northward it appears to rise sharply to a point; and when with great circumspection I crawled up I found myself on a granite summit, not three feet wide. I was rather mortified to find that the most westerly and another peak to the east appeared higher than where I sat, but certainly not by more than a hundred feet." As a matter of fact, St. John's Peak is only 15 feet lower than the actual summit. The central abyss at our feet, in the heart of the mountain, or at least a part of it, seems to have become included in the rather feeble term 'Low's Gully,' which St. John gave to the ravine by which Low reached the gap between the two pinnacles, and from which he looked down into the chasm below. The intended use of the name 'Low's Gully' may be referred to on page 281 (Vol. 1) of *Forests of the Far East.* For the terrific chasm itself, the name 'Low's Rift', 'Low's Chasm' or 'Low's Abyss' would be more appropriate.

One can well understand how the austere character of Kinabalu has impressed the simple-minded Dusun who live at its feet. Here are all the stern realities of nature—stupendous chasms, sheer cliffs, raging cataracts, and barren rock upraised hundreds of feet above the tree limit. It is only natural that its uncompromising aloofness should have suggested to the human mind an unconquerable remoteness such as that to which men's spirits retire after death. When at last we reached the

(*Upper*) THE CENTRAL ABYSS.
(*Lower*) THE ACTUAL SUMMIT OVERLOOKING THE ABYSS.

46

summit of that dreadful upland, the old priest pointed down into the abyss. "Fire both salutes in that direction" he said—and the noise of the discharge, feeble though it was, echoed back and forth amongst the precipice.

It is fascinating to speculate what the geology of Kinabalu can be. It is a vast mass of granite—naked and precipitous. Far below are ranges of shale and reddish laterite that once were clay; and here and there, scattered over North Borneo, are isolated fragments of limestone, always full of caves. From the order and situation of these rocks it seems we must picture a time when there was a limestone stratum, probably of great thickness. Below that was a deposit of clay. Then, possibly as the Burmese-Malayan Arc, the granites of Kinabalu burst through the clay and lime the thrust must have been terrific. St. John mentions that some of the strata near the base of the mountain are nearly "perpendicular, being at an angle of 80 degrees." Subsequently there may have been a further general rising of land, followed later by a period of great depression that plunged all the lower parts of Borneo into the sea. Then the land rose again, and has since been subject to an enormous denudation from rain, and from the swift descent of its short rivers. As a result, the soluble limestone has almost entirely disappeared. Only a fragment survives. The clay, too, has melted into comparatively low ranges of from 3000 to 5000 feet. Even the granite has suffered, as the precipitous walls of Kinabalu show, drenched as they are by constant rain that flows off quickly in countless waterspouts. The great waterfalls soon disappear after rain has ceased. But the granite is hard. It has stoutly resisted those age-long forces of frost and rain; and its head rises still to over 13,000 feet.

And the tale of depression and upheavals of Borneo, with all the tragedy they involved, is certainly indicated by the specialized fauna and flora of Kinabalu. Nothing can be more romantic, if one pauses to think of it, than that the frail insects and butterflies of the Malay Peninsula should exist almost in their entirety in Borneo. Even if their forms are slightly changed by the long isolation, their habits are identical. The insect inhabitants of the deep forest, the lovers of scrub or garden, are the same here as on the mainland of Asia, for they came here overland across country that has since sunk and become a shallow sea. But in spite of this, most of these species, as we have seen, have not invaded even the

47

base of Kinabalu, where many purely local and specialized types occur. Yet they, too, are Asiatic.

We must suppose then that there was originally an ancient Asiatic fauna that was largely wiped out of existence by the submergence of Borneo, or more probably by a succession of subsidences, in which Kinabalu alone remained above the sea to offer a refuge. There these types survived, and, by long isolation, have assumed specialized forms; and perhaps their alpine character has tended to keep them distinct from a more recent fauna which restocked Borneo when it rose again from the sea, and was rejoined for a time to the Malay Peninsula and Sumatra (but not to Java) in comparatively recent times.

It is fascinating, I say, to leave the imagination free to perpetrate catastrophies of this sort with a Continent and the second largest island in the world; for there is increasing evidence, including the mute evidence of such small folk as the Lepidoptera, to show that something of this kind did actually occur.

But Nature, even in the Tropics, will not permit a protracted indulgence in abstruse themes at 13,000 feet. We had already been over an hour at the summit. The cold, as soon as we ceased to move about, was intense. Ice has been reported, but we found none, perhaps on account of the recent rain. On our arrival at the top visibility was good. After admiring fearfully these peaks and precipices of Kinabalu, the eye naturally sought relief in the glorious panoramic view of North Borneo that lay stretched at our feet. Towards the sea spread Murudu Bay and Kudat, Usukan Bay, the islands of Jesselton, and the coast of Benoni and Brunei. Inland lay Bundu Tuan and the Renau Plain, and then range upon range of hills, flecked with tiny clouds; and even as we watched, the flecks grew and hid that fair landscape, so that in quite a few minutes we looked down upon a rising sea of clouds. Then just as suddenly mists came bowling down from above. A wind moaned among the peaks; and from the abyss white lines of wraiths came hurrying along the ridges, as if the ghostly inhabitants of Kinabalu were impatient to resume their age-long solitude.

We saw no life on the mountain-top, except one bird, that I noted in my journal. "It was as large as a crow, and had a yellow beak, but no more

WARNING SIGNALS. THE GATHERING OF THE MISTS.

could be seen at a distance." I have since presumed this bird to have been the "Blackbird having a golden bill" of Burbidge, which Whitehead, on page 191 says is *Merula seebohmi*. As a matter of fact, the number of essentially high-altitude birds, even in the forest zone at Paka Cave, is limited, and their scarcity may have some relation to the geological age of the mountain. It is true that tall, solitary peaks do not tend to evolve a large number of high-altitude birds. On Ruwenzori, I believe, there are only 95 zonal species. But the recent discovery of Kinabalu types on other Bornean highlands justifies one in regarding this peak, solitary though it is, as a part of the main upland system of the island in the ornithological sense; and if that is so, then the limitation of high-zone species suggests that the mountain is not of great age. In the Andes (a 'long' Tertiary range), according to Dr Frank Chapman, no less than one third of the bird life, excluding marine and migratory forms, is altitudinal—a far larger proportion, I imagine, than is the case on Kinabalu. At the bleak summit there were no butterflies about. Of plants, there was a flower like a buttercup growing in the cracks of the rock, and a silvery-leafed plant rather like Edelweiss, but with a yellow flower, which Whitehead says is *Potentilla*. Besides these, there were a few mosses, lichens and the dwarf *Leptospermum* I have already mentioned; and grasses that are the food of ghostly buffaloes.

The whorls of mist that now enveloped us were an urgent hint that it was time to quit the summit. There is only the one way down, and to miss it in fog would not be difficult. We therefore started the descent shortly after 10.00 a.m., reaching Paka Cave by noon in a state of exaltation, which we heightened with a liberal tot of whisky to commemorate our success. All the wet of the ascent, all the cold of the summit, were forgotten. There remained only the memory of our marvellous feats. Kinabalu had become a family epic.

Chapter VI

BUNDU TUAN

On the following morning we descended from the mountain. The porters now needed no urging. At Kamburonga (7000 feet) I caught a *Danais crowleyi* (Plate I, 9), which was the highest elevation at which a butterfly was seen. The observations of Whitehead, who spent six weeks on the ridge at Kamburonga, tend to show that at all times butterflies are scarce at this elevation; and it is interesting to note that the highest altitude butterfly he caught was at this very place, and was a *Danais.* He says, on the sixth day of his residence:—

"I caught to-day the first butterfly I have seen at this altitude, a species of *Danais;* but as the members of this genus are numerous and hard to separate into species, this butterfly is at present unnamed. The only other species of butterfly I saw at or above this camp was a blue comma *(Venessa perakana),* and a 'skipper' which was very dark, almost black, and which I failed to obtain." And on the eleventh day he writes:—

"The last few days have been most propitious for the appearance of insects, but I never saw a single butterfly, though I was out all day." The assumption is that butterflies are at least very scarce at the high altitudes. Passing our former camp at Lumu Lumu, we descended to Bundu Tuan in eight and a half hours. The last Expedition, that of Evans and Sarel, which had slept at the summit, made the whole descent in a day. No sooner had we regained the Bridle Path than we found ourselves once more in the thick of the butterflies.

Bundu Tuan (3340 feet) is a pretty Dusun village lying at the bottom of a valley at the base of the mountain. After our recent experience, its houses and fruit trees looked very refreshing and homely. Children were playing on the grassy slopes, and pigs rooting about under the houses; and from somewhere came the sound of music resembling a Yawyin melody. Even the dead, whose graves lay scattered about the village, seemed still to belong to it, especially one old fellow whose hat, coat and trousers had been set up like a scare-crow, and who seemed just to have risen from his sleeping place to stretch himself. In many places lay the half-exposed jars in which the Dusun finally bury their dead.

We remained resting a couple of days in Bundu Tuan, and a perfect rain of insects descended upon me. The most interesting acquisition was a mantis whose fore-legs ended in discs, as if the insect was playing a pair of cymbals. It is probably related to *Hestiasula sarawaca*, described on page 130 of Shelford's *Naturalist in Borneo*. The villagers were so anxious to earn 5-cent pieces, and took such an interest in my affairs, that it was impossible to keep them out of the rest-house. It was the first time that I felt really on intimate terms with these good-natured Dusun, whose character and appearance resemble those of our hill Kachins of Burma. On the second day a cow was ready, a great amount of wine supplied, and the porters had a splendid feast at my expense, in which half the village joined. They got very satisfactorily drunk; and next morning when I paid them off with a little more than what had been haggled for, many of them insisted on shaking hands with me before they left. The actual ascent of Kinabalu from the base to the top and down again, cost me $122; for it is another Kinabalu tradition that, though a hard bargain must be driven in self-defence, it shall not be taken advantage of.

The halt at Bundu Tuan enabled me to obtain several fine butterflies, particularly what appeared to be two very handsome species of *Delias*. As a matter of fact, both are *Delias eumolpe*, which is highly dimorphic—the male being white with a black tip (Plate I, 3), and the female, black with a bar on the fore-wing, and the inner half of the hind-wing white. In both sexes these butterflies vary much in size, and in both the reverse (Plate I, 30) is richly ornamented with black, red and yellow, the inner half of the hind-wing being entirely yellow. Another beautiful dark *Delias*, obtained at Tinompok, is *Parthenia*, which shows a little red on the upper side at the hinge of the hind-wing. It occurs nowhere else but on Mt. Kinabalu, and is uncommon there. I also got several pretty moths, including a Sphingidae, *Chaerocampa celerio*; a clear-winged species of *Syntomis,* with a blue, yellow-striped body; and a *Geometer* (unidentified), which is black, with one small red bar on the fore-wing, the hinges of both wings gleaming with a most lovely blue. An interesting catch was *Cyrestis maenalis seminigra* (Plate I, 28).

The dimorphic tendency of many Bornean and Malayan butterflies is striking, and is connected with the question of mimicry, of which I have written in my *Legend of Malay*. Here, amongst other butterflies, *Amnosia decora baluana, Lethe darena borneensis, Hebomoia glaucippe, Hypolymnas misippus, Pareronia valleria lutescens, Delias eumolpe*, and, in fact, all the butterflies 1 to 8, Plate I, are strongly dimorphic. In none of these cases do the males and females appear to be in the least related, so greatly do they differ from each other in size, shape, pattern and colour. In *Delias eumolpe* (Plate I, 3), species of each sex vary so much in size that large and small members of either sex would be mistaken for a male and female of the species by anyone not conversant with their dimorphism. This variation in the size of individuals is specially marked in Bornean males of *P. memnon sericapus* (Plate II, 5). I obtained one extraordinary little dwarf of *Cethosia* (Plate I, 19) that measured only two inches across, less than two-thirds the normal size (Plate I, 20). Such differences, however, are of no consequence from the point of view of classification.

When at last we had leisure to examine the collections, it was found that we had obtained 218 kinds of butterflies, of which 13 were new to me; 140 birds; 3 rats; 12 squirrels; 2 shrews; about 50 mosses, including one new species and one new variety of a species; uncounted insects, including beetles, weevils, stink-bugs, stick-insects, mantis, rhasinids, cicadas, locusts, fulgorids and trilobite larvae; and 24 species of spider belonging to ten genera and six families. Of these, two specimens may require new genera, and another two specimens are new species. The collection of spiders will, however, take some time to work out.

The question now arose of return to the coast. There were two alternatives. Either we could proceed further inland *via* Tambunan and Keningau to the rail-head at Tenom—this being the most attractive route; or we could return the way we came. I was particularly anxious not to attempt too much, preferring to see a little well rather than a great deal indifferently. Further, it would repay us to revisit the excellent butterfly grounds of which we now knew something. We therefore decided to return as we had come, by leisurely stages, reaching Kotabalud on the 26th June. There being no steamer due, it was necessary to travel thence overland to Jesselton. The 15 mile march over the hills to Tengilan was the hottest in my Bornean experience, and it required several bottles of beer in Mr Delap's hospitable house to restore me. On the following morning Delap and I continued the journey together. This was accomplished by boat through a mangrove swamp, with the writhing roots of trees rising on either hand from the lane of water. We had one distant glimpse of that Bornean curiosity, the Proboscis Monkey but I could not see them very clearly. Presently we came to a wide expanse of open water beyond which we landed at Tajau, and rode the remaining three miles into Tuaran. A twenty-mile motor drive and a couple of hours by train brought us to the sands of Benoni to rest, think, and write on the shores of a summer sea.

Chapter VII

BENONI

At Benoni I took a large, rambling house on the sea-shore. The railway line from Jesselton passes close behind it, and the train can be stopped by putting out a flag. Before the house is a belt of clear sand stretching for miles along the coast of Kimanis Bay. In one direction rises the hill of Papa, and in the other a belt of casuarina; and far away on the blue horizon lie the shadowy forms of three islands—'Pulo Tiga.' Benoni, as previously mentioned, was the *pied-à-terre* of the Kinabalu Expedition in Borneo, where we made our preparations on arrival, and to which we now returned, our work done, to revel in a wet, naked, unconventional life. For the first few days it was rather stormy, and the long lines of breakers afforded endless fun. Then a calm followed a lovely placid era of blue sea and blue sky, cool breezes and rich sunsets.

A calm sea at Benoni is not, however, ideal, since at such times the water becomes infested with jelly fish. They sting badly, and it is really not easy to avoid them. One of the hemispherical jelly fish proved to be of very special interest by reason of the little fish that attach themselves to it, make it their home, and live inside it. I believe that fish are known to live in the stomachs of sea anemones on the Barrier Reef, but I have never heard of their using jelly fish in the same way. These parasitic fish never go far from their living home, and do not seem to grow longer than about half an inch.

In the early mornings we sometimes saw little fish jumping frantically to an immense height, followed by a large, glittering fellow that I suppose was an Albicore. There are many wonderful things in these warm seas of which little is known. The most exciting and interesting is the Devil-fish—a great Ray—of which I heard many yarns. In spite of its bulk, it occasionally leaps clear of the water to shake off the Sucker Fish that attach themselves to it. The Saw Fish grows to an immense size. Its habit is to thrash sideways, so as to impale small fish on the teeth of its sword;

55

and when one of these monsters gets into the fishermen's nets it creates ruin and havoc. Darvel Bay is the home of the 'Dugong,' a sort of sea-cow, that has large breasts, and a small round head, and is quite uncannily human in appearance. From it is said to originate the legend of Mermaids. At Benoni, crocodile are occasionally seen in the sea, but they probably only travel without feeding from one river to another. The teeth which they shed from time to time are treated as charms by the natives. At Miri, in Sarawak, I saw the carapace of the gigantic King Crab—a formidable beast with a tail that it raised like that of a scorpion; and indeed, I believe that originally scorpions came out of the sea. The only crabs at Benoni are the little Pill Crabs, who scatter the sands with millions of pellets when the tide is out. I believe they extract nourishment from the sand by rolling it into balls.

In these seas the tides follow no known rules. The highest tides occur neither at the equinoxes nor at full moon. At Neap tides there are two daily tides, but at Spring tides there is only one; and occasionally there are four Spring tides in the month! In some places the maximum rise and fall is less than five feet, yet in others, as at the extreme east of Borneo where tides meet from two directions, the difference is as much as fourteen feet. These irregularities were very noticeable at Pangkor in the Straits of Malacca where I spent a holiday in 1924, and I supposed them to be caused by the narrowness of those waters. But here, at Benoni, the whole wide China Sea lay before us. However, the spaces between the islands of the Archipelago are not very great, and it is probable that the eccentricities of the tides are due to ocean currents which vary according to the Monsoon, so that the collection of data is of little value.

A curious feature of the China Sea, I am told, is a completely circular current, the cause of which is imperfectly understood. The existence of land-masses like the Philippines, Celebes, Borneo and Java profoundly influence the flow of winds. Were these islands not existing, there would be no Monsoons, but merely steady Trade Winds.

However, none of these things bothered us much in Benoni. It was essentially a period of rest, though the time was pleasantly employed in writing up accounts of the Expedition and in studying, describing, arranging, and as far as possible identifying, the collections of birds and butterflies.

Benoni is hardly as ideal as Pangkor in the Straits of Malacca, the *pied-à-terre* of my last leave. It has a special brand of sandfly which is more rapacious than all the Devil Fish that ever came out of the sea, and to which may be attributed a crop of nasty sores which we carried away from Borneo and nursed for the next two months. Besides, Kimanis Bay is large and featureless, without the individuality of Pangkor; and there are no fishermen and few shells. But the sands are magnificent, and we were very happy bathing twice a day, morning and evening; and sometimes again a third time in the warm moonlight.

Chapter VIII

INTERIOR

The only railway in British North Borneo runs—or perhaps it would be more truthful to say 'crawls'—along the coast from Jesselton and Benoni to Beaufort, a distance of 57 miles; and thence another 30 miles up the valley of the Padas River to Tenom, the headquarters of the Residency of the Interior. Until I went to Borneo I had never imagined a railway train could be so exasperating, could take so long to screw up courage for a start, or seize so readily any excuse to stop and take root in its tracks.

The last section, up the Padas Valley, affords some really beautiful scenery. From about miles 73 to 85, the river traverses a gorge or defile, its waters breaking into a succession of rapids as it comes tumbling down a boulderstrewn bed. High hills rise on either hand, clothed from the river bank to their summits in dense and splendid jungle.

The railway follows the right bank, and there are several impressive aspects of the gorge, especially if the journey is made, as we made it, by motor trolley. In wet weather the line is often blocked by serious landslides, and it is rather an expensive section to keep up. Where the rock is exposed, right close down by the river, and in a few rare cases where the upper slopes are too steep to hold earth and vegetation, the strata are seen to be tilted at an angle of 60 degrees, and to be in some places very highly contorted. The formation has received rough treatment at some period, possibly when the granites of Kinabalu smashed up the surface of Borneo. Later, there was probably a lake at Tenom above the gorge; and indeed the Muruts have a legend regarding such a lake, and about a giant who kicked down the gorge by which the Padas River traverses the range.

Having taken the motor trolley, we were able to stop for a couple of hours at Rayoh in the middle of the gorge and look at the butterflies; but for that matter there is probably no B.N.B.S.R. bye-law against chasing them from the train during its reckless career, or while the engine-driver is away cutting firewood. I should imagine the butterflies to be very interesting earlier in the day. As it was, in the early afternoon, we got *Brookiana* (Plate II, 1), and a *Leptocircus meges* (the only specimen obtained, though, as already mentioned, this butterfly occurs at Bundu Tuan). It is far less common than *curius*, and is never seen in clouds. There was also an abundance of the commoner Papilios and whites. The upper end of this gorge was the only place in Borneo in which *Euploea* were seen in any quantities; but here, on the dripping surface of a cutting, we came upon crowds of them, six species being taken in the net at one time, namely *E. diocletianus* (Plate I, 15); the shining blue *E. mulciber*; the handsome plain velvet-black *E. deione zonata* (Plate I, 13); two Blacks *E. aegyptus* (Plate I, 14); and *E. morrei*, with white spots on the wing-tips; and that charming little miniature *Euploea, Mezares aristotelis* (Plate I, 16). Their comparative scarcity near Kinabalu made them now the more acceptable to my eyes; and, indeed, there are really few butterflies so lovely as these common velvet-brown *Euploea* with their rich blue iridescence.

At the 85th mile the gorge ends, and the country opens out into a hill-locked plain, very green and beautiful—the bed, as I have suggested, of an ancient lake. The Padas River, which flows from Ulu Padas, makes a sharp bend here to force its passage through the hills, and is met on the plain by another river, the Pengalan, which rises near Tambunan.

At Tenom ends that romantic bridle-path that we followed from Usukan, across the lower spurs of Kinabalu to Bundu Tuan, and which we might have followed *via* Tambunan and Keningau to Tenom, had we not elected to return from the mountain the way we came. This end of the bridle-path also is pretty fair butterfly ground, and I spent my last hunting day in Borneo along it. Within a couple of miles from Tenom we saw the yellow-necked Ornithoptera, *Amphrysus flavicollis* (Plate II, 2 and 3) already mentioned; and also a male and female of what appears to be Ornithoptera *Helena cerberus*, with red on the neck and body. This is the only time a red-necked species was seen, and to my surprise the subsequent identification of both these splendid black-and-yellow Ornithoptera proved difficult, and was only established provisionally after transferring them from the Singapore Museum to Sarawak for reference. The male of the supposed *Helena cerberus* has the fore-wing almost plain black with only very faint splays; and the hind-wing gold. But Mr Moulton, to whom the specimens have been shown, notes that he has never seen anything of the kind before, and suggests they be referred to Europe. Naturally the possibility of adding a new species to the most splendid family of Ornithoptera is not a little exciting to a collector. The female of this supposed *Helena cerberus* is quite the handsomest of all the Ruficollis type, being even whiter on the fore-wing than the female *Amphrysus flavicollis*.

Tenom is quite a little place; indeed, hardly seems to justify the existence of a railway. There is a station, a playing ground, a row of Chinese shops which trade with the Muruts of the surrounding villages, and a low green hill on which is situated a flag-staff, and the hospitable house of the Resident, Mr Wooley. Tenom is 600 feet, and round about, at some distance from the plain, are encircling mountain ranges.

We are now in Murut country, the Murut being rather naked, simple children of the hills, not very different in essential matters from the Dusun, through whose uplands we travelled to Kinabalu. The Murut usually wear the hair long and knotted behind, and in a low fringe on the forehead. Their principal, and sometimes only, costume, is a short beatenbark *chawat,* or loin-cloth; though the Murut of Ulu Padas and other parts wear a linen loin-cloth that is quite 60 feet long. On swell occasions a khaki coat may be added, which, however, rather accentuates the absence of trousers. But no Murut is conscious of this deficiency, especially if he is wearing a felt 'homburg.'

The Resident himself took me over a Murut house—one of the few in the immediate vicinity. Tenom, though officially 'Interior,' is only the beginning of the real Interior. This house is 33 paces long, which is a fairly average length, though occasionally they are seen up to 120 and even 150 paces, forming the communal residence of a whole village. The one we have chosen to describe houses seven families. It is built of bamboo, roofed with palm-leaves, floored with *nibong* planks, and raised about six feet off the ground on a forest of piles. There is no front hall or gable, as in Kachin houses in Burma, but the place is entered through a door at one end by means of a notched log that is nearly perpendicular. How the dogs get up is a marvel. Before entering, we had to sweep off a puppy that was stranded on the second notch, howling miserably.

One side of the house is partitioned off into private sleeping and cooking quarters. These are usually in the form of cubicles for each family, though in this case there were no such divisions, but merely a number of fire-places. The other side of the house, down its entire length, is a vast common room. Above it is a sort of gallery in the roof where the bachelor lads can sleep when the place is crowded. On the whole, the building was remarkably clean, though the Murut cannot be accused of sanitation.

The interior of a Murut house is full of interesting things; the daily life of the people being epitomized by their various belongings. Along one wall of the common room are ranged fifteen or twenty large jars, some old, some comparatively new, and all probably of Chinese manufacture. Rice and wine are stored in these jars, some of which are glazed and quite handsome. A score of gongs hang along the wall. They constitute part of these people's wealth, and in the pawn-shop in Kotabalud I saw literally hundreds of gongs and brass pots on which money had been temporarily raised. Some of the gongs are rather deep, some shallow, and some of medium depth and shaped like Kachin gongs—though few have the exquisite tone of the gongs of Burma. They are usually made by the Malays of Brunei or Kimanis.

Then there is a vast array of articles made of grass, cane or rotang, some in use, others stored away in the roof, and others again in course of manufacture by the women. These articles include baskets, hats, traps, mats and winnowing-fans. The Murut are clever weavers, and have invented a number of interesting patterns for their mats, which are quite a science, and display considerable ingenuity and imagination. Some are a sort of hook pattern; others depict men and animals more or less conventionally. For instance, one design is meant to represent four men drinking from a jar; another four men with their heads on one pillow; and yet another four human heads hanging on a post. Such patterns are cleverly twisted to fit any required space without breaking the design. All these kinds have special names; and the edgings are another science in themselves. The most complicated pattern of all is well named the 'mad pattern.' The various races of Borneo do not have their own bags or haver-sacks, as is the custom in Burma; but clans, districts and even single villages have distinctive hats and baskets.

A deer trap was one of the most interesting things I saw in this house. It is made of rotang, in the form of a rope, which is really a portable fence in lengths of about 60 feet. The whole way along, the rope is furnished with large nooses, so close and over-lapping that when a deer runs into it, it is bound to become entangled in at least two or three loops. These deadly traps, into which deer are driven, are set up in the grass-lands beyond the river.

Other hunting appliances are rough spears with the blades bound to long shafts, and blow-pipes, which are made of wood (not of bamboo as on the Malay Peninsula), and furnished at the end with spear-heads. Unlike the bamboo pipes of the Sakai, they are rather heavy; yet, if correctly held, can be kept quite steady.

The war gear of the Muruts—one can hardly call it 'armour'—is largely made of wood or rotang. It is no doubt effective against the weapons used—notably the blowpipe dart. The shield is made of wood and is highly painted; the helmet of rotang or cane, with plumes. The body is protected with a sort of vest, either of woven cane, or of black bear-skin, studded over thickly with white cowries or shells. In addition, these vests may be ornamented with black and white porcupine quills, or with a layer of feathers from the hornbill. The Murut warrior in full paint cuts a brave figure.

The sword is slightly curved, the handle being double pointed like the open jaws of an alligator—though it does not represent that. These swords are only used in war, and not, as in the case of the dah of Burma, for daily and domestic purposes as well. The scabbard and hilt are ornamented with three or four bunches of human hair, and perhaps a couple of 'eyed' feathers of the beautiful Argus Pheasant.

In the roof of this house hung half-a-dozen human heads, each with a pig's tusk stuck in the aperture of the nose. The Murut keep the skull only, and not the mask as do the Dayaks. These heads, the householders naively told us, had been taken as recently as 1916! when, it appears, there was a small rising in these parts.

63

However, it must be understood that the essence of head-hunting is killing for the express purpose of securing a head. Merely to remove the head of a dead enemy is another matter.

Formerly all Murut houses were of the communal kind, and many were strongly fortified. But now that profound peace and security have settled upon Borneo, the tendency is to split up into small family houses. This habit is naturally deplored by the Chiefs and Elders, whose influence is thereby diminished.

No such 'long' houses were seen in the Dusun country, because, with a few exceptions, communal houses were never a Dusun custom.

Chapter IX

INDIGENOUS RACES OF BRITISH NORTH BORNEO

Dusun and Murut are the predominating tribes of British North Borneo, where they furnish nearly three-fifths of the population. Messrs Hose and M'Dougal class Dusun under the head of Murut, and believe the Murut to be immigrants from Annam. That these people are Proto-Malays, and have a Mongolian origin, and that they are related ethnologically to the innumerable hill races scattered over Cambodia, Siam, Burma and Assam seems probable, though definite proof of such connections is still lacking. There are, however, many points of resemblance. A similarity between the Murut of Borneo, the Abors of Assam, and the aborigines of Formosa has been noticed. The long communal houses, the method of hill cultivation, the rice wine of Borneo, are all repeated on the Burmese border. The Murut men wear the same bamboo garters, teeth necklaces, and hanging mats to sit upon, as the Chin Boks of Burma; and both Dusun and Kachin women wear hoops of lacquered cane round their bodies. The Murut mat, however, is oblong in shape, pointed at one end, and made of woven grass; and the hoops of the Dusun women are usually worn round the breasts instead of at the hips as in Burma. The Animism of the Dusun, with its spirits of trees and mountains, resembles that of the Kachins. As the Dusun dead go to Mount Kinabalu, and the Murut to the uplands of Mulundayoh and Antulai, so do the spirits of Kachins return to a mystic flat-topped hill called Majoi Shingra Bum, and those of the Sema, Lhota and Ao Nagas to the mountain called Wokha, that rises conspicuously out of their country. Dusun Chiefs are called *Orang Tua*; Kachin Chiefs have the title of *Duwa*. Both races have the same sort of character, the same superstitions, the same omens, the same curious belief in the sanctity of stone implements, of which the Bornean name (probably Bajau) is 'Gigi guntor'—Thunder Teeth. The Kachins call a thunder-bolt *Mu-ningwa*, which means 'Heavenly Axe-head.' The Head-hunting of Borneo has the same features as that of the Wa in the Shan States, and of the Sangtam Nagas in Assam. But curious and suggestive as all these things are, we are obliged to recognise that they are superficial; that as yet no connection has been scientifically established between the semi-wild races of Indo-China and those of the Malay Archipelago. The problem presents itself to me in the following form. There are traces on all sides of several extensive migrations in this part of the World and, as far as the geography of the land permitted, their general direction was to the south. In prehistoric times the first known purely human inhabitants of Java were an Australian type represented by the Wadjak man, and, as far as we can guess, they passed south-east about fifty thousand years ago and became the ancestors of the present Australian aborigine. In view of the known oscillation of land in this part of the Earth's surface, it may be noted that a rise of 600 feet would reduce the sea-journey to one of a hundred miles. Secondly, there seems to have been in prehistoric times a spread of Negroid types, whose characteristics are visible today in certain races of Southern India, in the Negritos of the Andamans, the Malay Peninsula, and the Philippines, and in the natives of Tasmania and the Pacific. Indeed, Wadjak man, though essentially Australian in type, has also much in common with Negroid types. He seems, in fact, to link the fossil Rhodesian man with the Australian aborigine.

Now, in historical times there has been for two millenniums a steady southward Mongolian migration into Indo-China—that is, into Assam, Burma, Siam and French Indo-China. This movement is not only historical, but is actually in progress, and may be observeded in many races of Burma such as Kachin, Chin, Shan, Miao, Shan Tayok, Lisu and others. The most southerly point to which this particular series of Mongolian waves has reached is marked by the Shans in the Isthmus of Kra, and by the Siamese (who are Shans) at Kroh in Upper Perak. That is to say, this great movement has not yet penetrated further than the extreme north of the Malay Peninsula.

Between these historic and prehistoric movements lie the Malays and Proto-Malays, whose origin is so perplexing. The first historic appearance of Malays is in Sumatra and Java. How did they get there? The existence of an age-long and all-compelling migration suggests that they too must have shared in it, and come south. Possibly their ancestors swept away an Australoid population from Java. Certainly they spread *north* from Sumatra and Java into the Malay Peninsula, but in this they followed a well authenticated rule that these races, while turning instinctively to the south, will indeed move north to occupy an empty space. Much of the Malay Peninsula was empty. The northward move of the Kadus of Burma is a prehistoric instance, and of the Kuki Chins an historic instance within our own times. Further, the Malays of Java and Sumatra moved south-east—that is, as directly south as the land permits—and we find them now scattered throughout the Malay Archipelago. There is a strong physical resemblance and a remote linguistic resemblance between Malays and Proto-Malays on the one hand, and between the Indo-Chinese races on the other; and besides this, there are suggestive parallels between their respective customs and legends such as I have indicated. The problem then resolves itself to this: From whence did the Malays reach Java and Sumatra? The suggestion is that they are a Mongoloid type from the north, who have participated at an early date in the general southward migration. But where is the proof? Possibly further research will some day supply the missing link. As far as we are here concerned, Dusun and Murut may be regarded as indigenous to Borneo.

Generally speaking, the Dusun occupy the plains and coastal ranges, while Murut occur further inland on the uplands of the Interior. Formerly the Dusun were referred to as 'Id'aan,' but the term is no longer used. 'Dusun' is a Malay word meaning 'Orchard,' suggested as a name for this people by the fruit trees planted round their villages. 'Murut' also is not an indigenous name. There are many subdivisions of these tribes, all having their own names. Even the word 'Dayak'—a corruption of *Orang Daya*, is a comprehensive term meaning 'Inland Men.' That all these Bornean races are related to each other is certain.

As for the Dusun, they have intermarried with Chinese settlers from a remote age, and have acquired certain Chinese customs and characteristics; but they are most certainly not of Chinese origin as some have asserted. There are three exploded myths in North Borneo: the suggested meanings of the word 'Kinabalu'; the reputed lake that lies beyond it; and the Chinese origin of the Dusun. I would not for the world revive these controversies; but with regard to the mythical lake, I would suggest that, since formerly the Dusun travelled even less than they do now, they may have mistaken a distant glimpse of ocean for a lake. Or there may have lingered the memory of lakes that once occupied plains such as those of Tenom and Ranau, and which may have survived till human times. Or possibly, as I have read somewhere, the reputed lake may have originated from the appearance of the Ranau plain in flood—the Malay word for 'lake' being *Danau*.

The Dusun, and the Murut also, are simple, attractive folk, as jungle-people often are. They are Animists, worshipping and fearing the more or less hostile spirits of mountains, rocks, and trees, and of their ancestors. Intensely superstitious, they have great faith in omens, particularly evil omens, as the crossing of one's path by a snake, scorpion, centipede or mouse-deer; or the hearing of the calls of certain birds. If such be seen or heard, a journey, marriage or other enterprise is abandoned. One of their most interesting superstitions, which seems to be peculiar to Borneo, is their veneration of certain porcelain jars of Chinese origin. These divine jars are of various grades of sanctity, the most sacred kind not numbering more than thirty. They are heirlooms of antiquity, in which the people hold part ownership; and there is a regular ritual for their worship—including an occasional debauch; for the jovial Dusun are frankly fond of a drunk.

It is probable that many jars now seen have been introduced in comparatively recent times by Chinese, who trade upon the local superstition. As the hill-tribes get civilized they lean towards Christianity or Islam, and the extraordinary value set on these sacred jars tend to depreciate. But certain gongs, too, are still highly prized; and some years ago, when one of these was confiscated and auctioned, it was bought in for $5,000.

As noted elsewhere, the Dusun suffer to some extent from skin diseases. There is, however, little goitre or venereal, though the latter is spreading. It is quite possible that the future prosperity of the country, which needs population more than anything else, lies in the proper handling of venereal, and in checking the high infant mortality that exists amongst these ignorant people. If average families have eight or ten children, of whom only two or three survive, it is obvious that there is a wastage in the very commodity—human beings—in which Borneo needs to be most economical. In Burma, where little is done in these essential matters, seven children out of every ten die before they are five years old! There is, however, this to be said for Burma in mitigation of such neglect, that as now 70 per cent. of the population is infected with venereal it is almost too late to do anything. A policy of one low-paid dresser and one midwife in every Dusun village, before conditions became as hopeless as they are in Burma, would probably pay hands down in the long run. In Bundu Tuan I was besieged with poor people who showed me horrid and suggestive sores.

Generally speaking, British North Borneo is administered under the Indian Penal Code; but in matters of inheritance, debt, and sexual crime, tribal customs are recognized. And, indeed, in cases of marriage where communal property such as gongs and jars are borrowed; and in cases of divorce where such property has to be restored, and the wealth later acquired divided, nothing but Dusun law could cope with the resulting confusion. As a rule, a Dusun has only one wife, but there is nothing to prevent polygamy, though the first wife, and in some cases her children also, have to be compensated. However, a pig or a couple of fowls will suffice.

The tribal law of morality is strict and narrow, particularly with regard to what the natives consider incest. A marriage between members of the same village necessitates a costly purification. In some tribes the marriage of second cousins is prohibited, and in others necessitates a payment of fines. In old days, in cases of proved adultery, the woman and her lover were tied together face to face, the man on top; and the injured husband was given a spear which he could drive in as far as he liked—that is, he could spare his wife if he chose. But, it is said, he seldom did spare her.

The head-hunting raids of former days have been entirely stopped, but the instinct to hunt is merely dormant, and frequently breaks out in individual cases. It usually takes the form of a vendetta—that is to say, there is some motive other than that of securing a head; and in this there is at least some mitigation. Three or four of these cases occurred while I was in the country; but near Tenom there was also a real head-taking case, the first for several years. In this instance a solitary Chinaman approached a Murut's hut in the jungle, apparently to ask for food. The opportunity appearing favourable, the Murut killed him to secure the head. Unfortunately for himself, the Murut boasted in his cups. Head-hunting is, in fact, becoming less attractive on account of the secrecy that now has to be maintained. The brave warrior can no longer flourish the head about for the girls to admire, for publicity inevitably leads to detection—and retribution. However, incidents are bound to occur for another generation or two, for head-hunting was ingrained into the very character of the people. To take a head, whether of a babe or a warrior, was a sign of manhood; and no maid would look at a boy till he had secured his trophy. It is not so long ago that a legal-minded Dusun came to the Civil Officer for a head-hunting *license!*

The Kinabalu trip was through the Dusun hills. Tenom in the Interior is Murut country. Owen Rutter says of the Murut that:—

"They number about 28,500, and are more or less naked, except for a loin-cloth; which, however, is often of great length. Formerly they wore beaten bark. Lowland Murut build comparatively small houses, but in the hills they still have their communal buildings, often 200 feet long, and for choice built on some knife-edged hill for protection from attack. These people are frightfully dirty, and suffer from ulcers, sore eyes and skin diseases. There is much blindness amongst them. Infant mortality is high. Venereal diseases, introduced by Dayak and Arab traders, have spread disastrously. . .

"Notwithstanding his little idiosyncracies, the up-country Murut is perhaps the most likeable native of North Borneo, where it may be said that the further one goes from civilization the pleasanter are the people met. It is doubtful if any native of any country in the world is so easy to get on with as the Murut in his normal surroundings, though admittedly he is not such an asset to the country as the more prosperous Dusun, nor is he so encouraging from the administrative point of view. He is a primitive animal, hardly touched by the outside world. He is hospitable, good humoured and honest; so honest that theft is almost unknown, except the occasional theft of someone else's wife."

No one seems to love the Bajau. Living near the coast. he is , more or less civilized, but appears to have derived little moral benefit from that. Nor is it altogether surprising for blood *will* tell; and the Bajau are simply the descendants of pirates and sea-gypsies from Sulu. They are, however, well spoken of by the timber firms who largely employ them; and in the water thay are as thoroughly at home as the fish.

I saw little of the Ilanun, though a community of them occurs near Usukan. They also were amongst the worst pirates in the old days. Their original home was in the Philippines. They and the Bajau are bold horsemen, and ride down deer and pig which they spear with extraordinary skill. Their sturdy little 14.2 ponies are said to have been introduced from Sulu, and remind one of what the 'Burma pony' was a decade ago, before he was spoiled.

The Kadayan are a race found scattered all along the coast of Sarawak and British North Borneo. The legend is that they originate from a Javanese military expedition which was wrecked off Brunei and hospitably treated. They inter-married with Brunei Malays and Murut, and are now semi-pagan, semi-Mahommedan.

Of the non-indigenous races in Borneo, the Indian is as much beloved as he is elsewhere—not more! It is to be feared that Satan (in whom I still firmly believe) continues to find naughty work for idle hands. There are places in Malaya where the people, like the Beaver, have "turned unaccountably shy" of the Aryan Brother. In Taiping, shortly after my return, the Indian community towed one of its fraternity a couple of miles along the high-way behind a hired car; and, appalled at the cats' meat that remained at the end of the rope, left it in the road, merely advising the Chinese driver not to mention it. It would be hard to devise a way of murder less discreet. The Indian alone, of all the races who make fortunes in these parts, fails to raise his standard of living. There are many *Darwans* (watchmen) who have built up immense bank credits, but who are still *Darwans*. The *DudhWallah* (Milkman), who brings you a pint of potential typhoid on his cycle, may be worth ten thousand dollars, presumably because he has not had the large-mindedness to spend, though gifted in no ordinary degree with the small-mindedness to turn one cent into ten. The Indian invariably does police-work where it is still open to him, keeps his inevitable cows, and lends money when he can. But the indigenous people are now lamentably protected. In Borneo, land may not be mortgaged; nor can it be even sub-leased without sanction. This is all very hard and very wrong, but it is at least surprising that the politicians, say in Borneo, who waste incredible time frothing at the mouth over trifles, do not turn their great minds towards a more vital problem of self-preservation.

By far the largest non-indigenous community, of course, are the Chinese. They have probably been in Borneo for centuries, and have done an immense amount towards developing the country. The Chinaman, here, as always, is thrifty, honest, diligent and sagacious. Every effort is made to attract Chinese to the country in larger numbers; and possibly they might come over more readily if their language and customs were better studied than they are. The Chinaman, never getting justice at home, is extremely sensitive abroad if his interests are not fostered.

The Chinese in British North Borneo are largely Hakas and Cantonese, who are rather closely allied. Much of the business of Jesselton is in their capable hands, and they also take up land and cultivate it. Wherever openings occur, the Chinaman is ready to avail himself of opportunities. Provided he has a favourable start, the Chinese immigrant will usually make good. As in the Malay Peninsula, the Hailams have the monopoly of domestic service. In British North Borneo Malays are not numerous, but, of course, their language is in general use here, as throughout the whole Archipelago. According to the census for 1921 the population is 257,344, of whom nearly 80 percent are natives of Borneo. The Chinese, who exceed 37,000, are the only considerable foreign community. Of the total population, 94 per cent. of males and 99 percent of females are illiterate.

CONCLUSION

British North Borneo is still in its infancy. With regard to development, there has merely been a scratching on the coast. The Interior has hardly been touched. There are as yet no roads. The existing 90 miles of railway runs nothing better than goods-passenger trains, and these only once a day on the main line from Jesselton to Beaufort, and thrice a week on to Tenom in the Interior. With few exceptions, Europeans' houses are of wood, and roofed with bamboo thatching. In short, the country is still in the pioneer stage. There can be little doubt but that, agriculturally, there is a future before it. The climate is fairly healthy and pleasant. Amenities will follow in due course. The country's supreme need at present is money and population.

The tourist might—certainly would—cry out with an exceeding bitter cry against the accommodation and communications. British North Borneo is (fortunately) not yet the resort of globe trotters. Nevertheless, it is a grand holiday ground, a manly country for those who love the hills and forests, who revel in freedom, and who have an appetite for the strange and curious aspects of nature. For such, the prospect of 'development', and the inevitable disturbance of old customs, has no great attraction. For, as it is, North Borneo is a paradise for the naturalist who will delight in its flora and fauna; for the traveller who will be absorbed in its geography and ethnology; for the climber and the artist who will find all they need in the mysterious uplands of Kinabalu. For all these, the country offers unusual interests, and every comfort and convenience that is essential.

For British North Borneo is pervaded by a homely, kindly atmosphere, that assures the stranger of help and hospitality, and the best that can be done for him. There can be no question but that the European residents love this land of their adoption, and believe in it; and that they are only too anxious to show it as it should be seen.

And something of this same friendliness characterizes the natives too. Most of them are simple, pleasant folk; sometimes ignorant, sometimes dirty, but always cheery, willing and hospitable, having within them the milk of human kindness. Before them, too, is a future, when the country's development shall have raised them above their present ignorance and superstition. One does not ardently desire that these good people should be lifted out of happiness into that state that we fondly call civilization. Hospitals, roads and perhaps a little elementary education is all they need, and contact with the right type of British Tuan.

In Kinabalu, North Borneo possesses a unique attraction. This wonderful and mystic mountain, trod as yet by a paltry score of travellers, still withholds half its secrets. And these await the research of the curious and adventurous. As Moulton has pointed out, the crest of Kinabalu "is divided into an eastern plateau summit and western ridge summit; the former accessible, the latter apparently not." A summit has a strange fascination; but in future the exploration of other parts of this mountain will be more useful and will confer greater honours. It may be that new birds, new flowers, new butterflies exist upon those unknown slopes. A dozen virgin peaks await a conqueror, but so also do the chasms and valleys that have never been invaded yet by man. From the beginning, Kinabalu has been an object of awe and wonder—and so it still remains.

IN MEMORIAM

As this book goes to press, it is with grief that I have to record the death of Sau Nan, to whom the work is dedicated.

For many years in Burma, the Malay Peninsula, and Borneo, Sau Nan has been the gay and resourceful companion of my adventures, of which the last was this ascent of Kinabalu. Brave, loyal, simple-hearted he was, and absolutely true and incorruptible. In Burma he will be remembered for work done in connection with the introduction of the Kachins into the regular Indian Army. For these services he received a Sword of Honour from the Government of Burma. But the reward he most appreciated was the conviction that through his labour his countrymen were uplifted and enlightened.

There is devotion—and such was Sau Nan's to me—that touches the heart beyond power of expression. In fine weather he was gay and debonair; when trouble came he was in the thick of it winning through. He was essentially one of Nature's gentlemen; and spiritually he was uplifted by a living faith in Christ. I recall him now, kneeling up in the bitter cold of Paka Cave, praying for fair weather.

Farewell, Sau Nan, brave and trusted friend. I grieve for you as for a brother. Farewell! Farewell! Enter thou into the Kingdom of the Lord.

APPENDICES

APPENDIX 1

ITINERARY

1925

28th May	Left Taiping (F.M.S.) on two months' leave.
29th May	Reached Singapore.
30th May	Halted Singapore.
31st May	Left Singapore on *S.S. Darvel.*
1st June	At Sea.
2nd June	Off the Sarawak coast.
3rd June	Landed at Miri, in Sarawak, and re-embarked.
4th June	Reached Labuan and Jesselton.
5th June	Jesselton.
6th June	Jesselton to Membakut by rail.
7th June	Membakut to Benoni by rail.
8th June	Benoni.
9th June	Benoni to Jesselton by rail.
10th June	Jesselton to Usukan Bay by sea,and on 7 miles to Kotabalud.
11th June	Halted Kotabalud. 300 feet.
12th June	Kotabalud to Tamu Darat. 8 miles. 500 feet.
13th June	Tamu Darat to Kayabau. 11 miles. 800 feet.
14th June	Kabayau to Kaung. 11 miles. 1100 feet.
15th June	Kaung to Dallas. 9 miles. 2500 feet.
16th June	Halted Dallas.
17th June	Dallas to Tinompok. 9 miles. 4900 feet.
18th June	Tinompok to Lumu Lumu. 3 hours. 5500 feet.
19th June	Lumu Lumu to Paka Cave. 5 hours. 9790 feet.
20th June	Paka Cave to summit of Kinabalu. 2$^1/_2$ hours. 13,465 feet; and back.
21st June	Paka Cave to Bundu Tuan. 8 hours. 3340 feet.
22nd June	Halted Bundu Tuan.
23rd June	Bundu Tuan to Kaung. 21 miles.
21th June	Kaung to Kabayau. 11 miles.
22th June	Halted Kabayau.
26th June	Kabayau to Kotabalud. 19 miles.

27th June	Kotabalud to Tenghilan. 15 miles.
28th June	Tenghilan to Tajau by boat, to Tuaran by pony, to Jesseltonby car, to Benoni by rail.
29th June– 6th July	Benoni.
7th July	Benoni to Tenom by rail.
8th July	Halted Tenom.
9th July	Tenom to Benoni by rail.
l0th–12th July	Benoni
13th July	To Jesselton by rail.
14th July	Embarked on *S.S. Darvel.*
15th July	Left Jesselton. Reached Labuan.
16th July	Left Labuan.
17th–19th July	At Sea.
20th July	Reached Singapore.
23rd July	Left Singapore.
24th–25th July	Kuala Lumpur.
26th July	Sungei Siput.
27th July	Taiping.

APPENDIX II

BOTANY OF MOUNT KINABALU

Without training in Botany it is extremely difficult to give a coherent description of what we saw of the flora of Mount Kinabalu. Botany was an amateurish side-line, only attempted in view of the unique interest of the mountain. I am under obligation to Messrs Holttum & Henderson, of the Botanical Gardens, Singapore, for their patient identification of the collection, and their notes which have made it possible for me to compile a brief statement of our observations.

As explained in the text, the ascent of Kinabalu was accomplished in three stages:—the first to Lumu Lumu, 5500 feet; the second to Kamburonga, 7000 feet, and on to Paka Cave, 9790 feet; the third to the tree limit at Sayat Sayat, about 10,500 feet, and on to the summit at 13,455 feet.

The vegetation of Lumu Lumu and the lower levels generally consists of dense, sunless forests, the prevailing impression of which is of ferns, and of spongy, sodden growths of moss, that enshroud the trees, the rocks and the ground in one unbroken robe of greenery. The atmosphere of these dim jungles is depressing; but the mosses would repay careful investigation, since the variety of them is very great. Some of the most curious and beautiful are species of *Hypnodendron*, which grow on stalks that give these mosses the appearance of tiny trees. Of these a new variety *latifolium* was found of *Hypnodendron copelandii* Broth. The species has hitherto been known only from the Philippines.

Amongst the few people who have visited Kinabalu, there have been expert plant and bird collectors, so that probably the products of this portion of the mountain have been pretty well explored in these directions. The most interesting fern found by our Expedition was *Polypodium triangulare* Scort, which Mr Holttum says is decribed from Perak by Scortechini, and is known from Penang, and has possibly been reported under another name from the Philippines, but has not previously been recorded from Borneo.

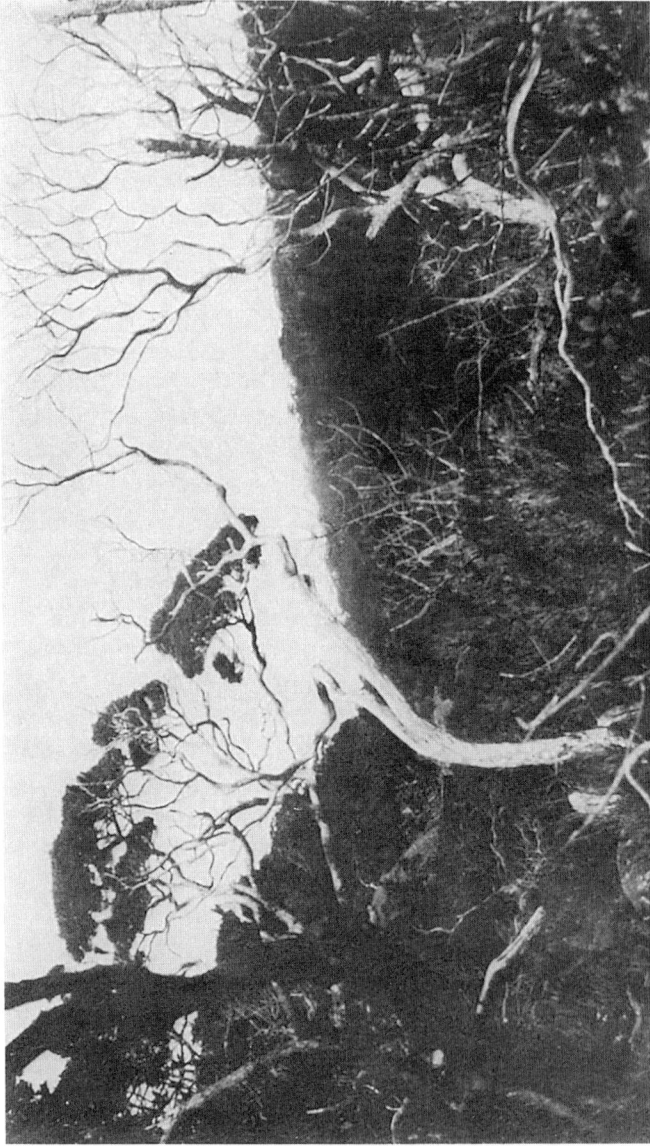

LEPTOSPERMUM RECURVUM, AT TREE-LIMIT, 10,500 FEET.

82

Above Lumu Lumu the ascent is steeper. Many orchids occur, and a salmon-pink *Rhododendron*. And at Kamburonga (7040 feet) a sudden change in vegetation occurs. The most noticeable trees are the bent and twisted *Leptospermum*, which here corresponds to the Juniper Zone of the Himalayas. Even more striking is the appearance of large and specialized *Nepenthes*, or pitcher-plants—namely *N. Iowii* and *N. villosa*. The *Nepenthes* have been discussed and described in some detail in the text (pages 33 to 35), and it is unnecessary to say anything further here about these extraordinary carnivorous plants, except that *N. Iowii* and *N. villosa* seem to be the only giant forms on this part of the mountain—*N. rajah* and *N. edwardsiana* occuring at their best on the Marei Parei Plateau.

From Kamburonga (7040 feet) to the tree-limit (10,500 feet) the predominating flora consists of two species of *Myrtaceae*— *Leptospermum recurvum* Hook, and *L. flavescens*, whose picturesque and twisted limbs are the key-note of the vegetation. To begin with, these curious trees are fairly large, but dwindle in size with the increasing altitude, until, near the extreme summit, *L. recurvum* occurs, along with a few other plants, as a dwarf hardly six inches high. This species appears to vary extraordinarily in form and habit with the elevation; the size, shape and leaves of the comparatively large trees at Kamburonga (7000 feet) being hardly recognizable as the dwarfs in the rock crevices of the summit (13,000 feet). Such dimorphism is, I believe, not uncommon in plants. Though *Leptospermum recurvum* is the predominating tree at the higher altitudes, and the most conspicuous by reason of its bent and tortured appearance, it is pretty liberally interspersed with rhododendron, and at least two species of dwarf pines *(Dncrydilum)* which are seldom more than four or five feet high, as far as I remember.

As already mentioned, salmon-pink rhododendron, *R. stenophymum* Hook., begin below Kamburonga, but above give place to two pinker or redder varieties, *R. durionifolium* Becc., and *R. rugosum* Low.

While the sudden appearance of *Leptospermum* is so striking at Kamburonga, the frequent mention of *Casuarina* at this very point by several reliable botanists is mysterious. There are certainly no *Casuarina* there now, as Miss Gibbs points out; and besides, *Casuarina* is essentially a coastal tree. Below Kamburonga occur a few *Lycopodium casuarinoides*, but they are not at all conspicuous.

Above Paka Cave the rhododendron are especially fine; and amongst them we saw one or two bushes with the beautiful white bloom of *Schima brevifolia*.

Generally speaking, the tree limit is fairly abrupt at Sayat Sayat at about 10,500 feet; but far above that, and indeed to within a few feet of the very summit, there are plants in the cracks of the granite, though the appearance of the mountain is of an uncompromising nakedness, except for a few copses of scraggy *Leptospermum* in sheltered ravines along the inner edge of the 'plateau.' The flora of the rock-crevices includes, as already mentioned, dwarf *Leptospermum recurvm* only a few inches high, and some lichens, grasses, *Potentilla* and Scrophulariaceae. The whitish lichens are probably *Cladonia* species, but remain unidentified. The *Potentilla leuconota* D. Don var. *borneesis* Stapf has silvery leaves and a pretty little yellow 'buttercup' flower. The Scrophulariaceae is *Euphrasia borneensis* Stapf, which has a small white flower touched at the throat with mauve. It has been recorded from the Philippines.

The following is a list of mosses collected. It will be noticed that they include one new species, and one new variety of a species hitherto known only from the Philippines:—

Thuidium plumulosum (Doz. & Molk.) Bry. Jav.
Leucobryum javense (Brid.) Mitt.
Dicranoloma angustifrondum Dixon sp. nov.
Homaliodendron microdendron (Mont.) Fleisch.
Floribundaria floribunda (Doz. & Molk.) Fl.
Trismegistia brauniana (Bry. jav.) Fl.
Mniodendron Mittenii Salm.
Mastopoma uncinifolium (Broth.) Card.
Dicranoloma Blumii (Nees) Par.
Hyphodendron arborescens (Mitt.) Lindb.
Hypnodendron Copelandii Broth. nov. var. *latifolium* (Distr. of type: Philippines)

APPENDIX III

SPIDERS OF KINABALU

Preliminary Report by Mr H.C.Abraham

TAIPING,
17th September 1925

Dear Major Enriquez,

We have just finished the preliminary examination of your Kinabalu collection of spiders, and as far as we are at present able to say, there are represented 24 species, which are distributed between 16 genera and six families. Of these, we think that two specimens will require new genera, and another two specimens are new species. The specimen from Paka Cave is not a true spider, but an *Opilione*, a sub-order with which I am not familiar.

The following is a detail list of specimens examined:—

Family	DIPLURIDAE	Genus	*Macrothele* (2 species).
	THERAPHOSIDAE		*Coremiocnemis* (1 species ? new).
	ARGIOPIDAE		*Araneus* (2 species).
			Argiope (2 species).
			Clitaetra (1 species ? new).
			Cyrtophora (1 species).
			Gasteracantha (3 species).
			Herennia (1 species).
			Leucauge (3 species).
			Nephila (2 species).
			Poltys (1 species).
	ATTIDAE		? new genus (1 species).
			? new genus (1 species).
	HETEROPODIDAE		*Heteropoda* (1 species).
			Panaretus (1 species).
			Theridion (1 species).
	THERIDIDAE		1 species.

Our idea is that Sworder shall take the collection to Europe with him when he goes on leave in January next, and compare them with specimens in the British Museum and, if necessary, Paris Museum. When that has been done we shall be able to produce a very useful paper about them.

Yours sincerely,
H.C. Abraham

APPENDIX IV

BIRDS OBTAINED BY THE KINABALU EXPEDITION

The Coast

Altogether 140 birds were obtained for Raffles' Museum (Singapore) during the approach to Kinabalu, and on the mountain itself up to Paka Cave (9790 feet). Above that only one was seen, and that distantly. I must again express my gratitude to Mr Chasen for the minute examination made of our bird and butterfly collections in Singapore, and for the care he took over identification.

A good many birds were collected on the sea coast at Benoni, and of these the finest were two Cuckoos. *Aegithina tiphia viridis* is grey on the head, throat and breast, with an orangered eye-patch, and a dull jade beak. The back and wings, and the long tail, are rich, dark blue-green, with the tail-feathers tipped with white. *Rhopodypes sumatranus* is almost identical with the above, but in addition has some conspicuous white marks on the back. Both are very beautiful birds. The Night-jar is common on the sand, and the gay little Iora in the thickets. Of several Kingfishers, perhaps the commonest is a lovely blue-green bird, with a white breast and collar—*Halcyon chloris cyanescens*. As noted in the text, this Kingfisher flew aboard the steamer in an exhausted condition while we were still out of sight of land off the Sarawak coast. *Halycon coromanda minor* is a rich mauve Kingfisher with chestnut under-parts, and scarlet bill and feet; while the largest of them all is *Ramphalcyon capensis javana*, a splendid buff fellow with blue wings and tail, and deep-red bill and feet. Two pretty little Sun-birds occur—*Aethopyga siparaja*, with scarlet throat and back, and a shining blue crown and cheeks; and *Anthreptes malaccensis* which is a larger bird with iridescent green head and back, touched with purple and blue, and turning creamy-brown at the throat. The breast and lower parts are bright yellow; the wings brown, touched with green and bronze; and the tail blue. This is one of the most charming of all Sun-birds. Its pear-shaped nest, made of fibre and lined with down, is hung at the end of a branch.

One specimen was obtained of the handsome Roller, *Eurystomus orientalis*, which has a dull greeny-brown head, touched at the throat with exquisite purple-blue. The upper feathers of the wing are bluish-green, the primaries black edged with Oxford-blue, and some of them barred with sea-green. The beak and feet are red.

Amongst the many other conspicuous birds of the Benoni coast must be mentioned *Sitta frontalis corallipes*, a small bluish Nuthatch with pale lilac under-parts and orange beak and feet; *Arachnothera chrysogenys*, a green spider-hunter; *Diopium javanense borneoensis*, a gaudy Woodpecker with scarlet crest, ocherous back, and mottled breast; a species of *Cyornis*, one of the blue, ruddybreasted Robins; and a blue-black hill form of Shama, *Kittacincla malabarica stricklandi*, with a white cap and tawny belly. Last, but very far from least, is the chestnut-red Pigeon, *Treron gulvicollis baramensis*, with black and yellow wings, a form with quite a restricted range; and *Conurus longicauda*, a Paraquet. Parrots are not very numerous in Borneo. This beauty has a dark green cap, with rosepink face, cheeks and nape, and with a black band at the throat tapering away on the shoulder. The upper mandible is red, the lower blackish. The belly is bright olivegreen, the back sea-green, the upper wing covets grass-green, and the outer webs of the primaries blue and green. From the tail spring two blue streamers of such a great length that they over-tip the rest of the tail by $6\frac{1}{2}$ inches—the whole length of the bird being $16\frac{1}{2}$ inches. At Benoni, where we spent several days, quite a large collection was made.

Kabayau to Kaung

As with the butterflies, so also with the birds, they were in the greatest profusion in the vicinity of Kabayau and Kaung at an elevation of from 800 to 1100 feet. They were still plentiful up to 4900 feet; but after that, and on the upper slopes of Kinabalu, were rarely seen at all.

At Kabayau we obtained a lovely 'Redbearded Green Bee-eater', *Nyctiornis amicta*, with a glorious lilac cap. The throat and breast are rich scarlet; the beak and feet black; the feathers of the tail are green above, but their under-surface is old-gold, tipped with black. *Rhinortha chlorophaea fuscigularis* a Malkoha (which was obtained also on the coast at Benoni) is grey on the head, neck and breast, turning to rich chestnut on the wings; and the long chestnut tail is tipped with white; the beak is dull jade-green. Three Woodpeckers were seen:—*Picus puniceus observandus*, which is green with a red cap, and red on the outer web of some of the wingfeathers; *Blythipicus rubiginosus*, which is sooty-brown with dull on the back and wings; and a brown Woodpecker with the wings barred with black that was too shot about to preserve. Specimens were also procured of *Copsychus saulari niger*, the mountain form of the common Straits Robin; *Aethostoma rostratum witmeri*, a Babbler; *Anthreptes macularia intensior*; and a Scimitar Babbler, *Pomatorphinus montanus borneensis*.

At about 1100 feet, near Kaung, two magnificent Broadbills were obtained. Of these the 'Black-and-Yellow Broadbill' *(Eurylaemus ochromelus kalamantan)* has a heavy, brilliant turquoise-blue beak. The breast is a lovely crushed-strawberry colour, though there is less of this than in the Malay Peninsula form. The head, body and wings are black, with a black necklace, and with touches of yellow on the back. The other, the 'Black-and-Red Broadbill' *(Cymborhynchus macrorhynchus)*, has the whole upper plumage black, with a broad crimson collar round the throat, and running back to below the eye. The breast, belly, and rump are also rich red, the tail black, and the inner wing-feathers white. The eyes are emerald-green, the feet cobalt-blue, and there is a touch of orange on the shoulder of the wing. The mandibles are turquoise-blue, with some yellow on the lower one. In stuffed specimens these brilliant colourings of the beak, which are so characteristic of these beautiful and curious birds, soon turn black. The 'Black-and-Red Broadbill' also occurs on the coast at Benoni.

An attractive little bird here seen is the emerald ' Blue-whiskered Green Bulbul,' *Chloropsis cyanlopogon*, which has a black throat, touched at the sides with cobalt-blue. There are also two handsome Cuck;oos. Of these *Zanclostomus javanicus pallidus* is dark grey with metallic-blue reflections. The throat, breast and belly are tawny; and the long grey-blue tail feathers are tipped with white. The beak is red, and the feet are black. In *Poenicophaes curvirostris borneensis*, the whole upper plumage is uniform deep blue-green, turning at the end of the tail to deep maroon. The breast and belly are also maroon, turning dull greenish at the vent. The upper mandible is jadegreen, magenta at the base; and the lower one magenta. The eye-patch is red. This is really the handsomest of all the Cuckoos we saw.

One female specimen was obtained of the Malkoha, *Rhinortha chlorophaea fuscigularis*, of which the head, back and breast are light chestnut, turning darker and richer on the wings. The long tail is black, indistinctly barred with dark-grey, and tipped with white. The beak is dull green. This also, in its way, is a striking bird.

Dallas

The altitude of Dallas is about 2500 feet, and the hills are sadly denuded of forest. Here we obtained the lovely sea-green Roller, *Cissa chinensis minor*, with a bright scarlet beak, and a black band running through the eyes and round the nape. The wings are chestnut; and the tail is pale green, broken on the under feathers with a bar. The feet are dull orange. *Rhinocichla mitrata trencheri*, 'Treacher's Red-headed Laughing-Thrush' is extremely common—a dark-grey bird with a deep chestnut head, and a dull orange beak. Of the many unobtrusive little birds that occur in such country, specimens were obtained of a Drongo, *Dicrurus leucophaeus stigmntops*; *Brachypodius atriceps*; and a Flycatcher, *Abrornis superciliaris schwaneri*.

Bundu Tuan and Tinompok

Bundu Tuan (3340 feet) lies at the bottom of a valley at the extreme base of Mount Kinabalu. Birds are naturally most plentiful at these lower levels, especially at Tinompok, which has an elevation of 4900 feet. Here three specimens were obtained of the Rain Bird, *Dendrocitta occipitalis cinerascens*. The crown and back are grey, the wings black, and the breast greenish-grey, turning dull buff lower down. The tail is black, except for two long grey feathers which protrude a long way out behind, and end in two dark blobs, rather like the floating tails of the more common Racquet-tailed Drongo. In some cases the lower part of these protruding quills are thinly webbed, but apparently only because they are old and worn.

A handsome Green Barbet, seen both at Bundu Tuan and Tinompok, is *Chotorhea monticola*. The crown is slightly bluish, decorated with a few touches of red. The beak and feet are black. But all our specimens were of young, immature birds.

Criniger gutturalis ruficrissus, a dullgreen Bulbul with a white throat, and dull chestnut tail, was seen both at Tinompok and Dallas. At the lower elevation of Bundu Tuan we obtained the charming little Sun-bird, *Leptocoma jugularis ornata*. Its upper plumage is dull green, with an iridescent purple patch at the throat and breast. The belly is orange; the beak and feet are black.

Lumu Lumu

One of the most interesting of all the birds obtained by the Expedition was a female of the specialized Kinabalu Trogon, *Pyrotrogon whiteheadi*, described by Whitehead as *Harpactes whiteheadi*. The beak and cheeks are bluish; the back is chestnut; and the lower back closely pencilled with black and brown. The wings are black, with the outer edge of each feather finely margined with white; and with the base of the inner webs also white. Some of the tail feathers end in black, others in white, and others again in chestnut. The breast is grey, the belly chestnut. It is not a very showy bird. The gaudy male, which is figured by Whitehead, was not seen.

Paka Cave

At the higher elevations about the commonest bird is 'Whitehead's White-eye,' *Chlorocharis emiliae*, a little green fellow with a dark crown. The preparation of specimens in the wet and cold of the ascent was persevered with, but under difficulties. Shooting was hard on account of the labouring of the breath. Specimens, even when shot close up, often fell a long way down those steep slopes, and a good many were lost in the undergrowth. On the whole, the birds were shy, and far from confidential, as St. John and Whitehead found them in 1858 and 1887; and this is curious, since they can have been little disturbed in the interval. The weather may have had something to do with it.

At the time, I was much disappointed that we did not get more than three kinds of birds at Paka Cave (9790 feet). But I find that in this connection Whitehead, the best authority on Kinabalu birds, says:—

"The species inhabiting the region above 8000 feet, and which do not descend much below that altitude, are *Cryptolopha trivirgata, Oreoctistes leucops, Androphilus accentor, Corythocichla crassa* and *Cuculus poliocephalus.* I only noticed three species above 10,000 feet— *Merula seebolInti, Cettia oreophila* and *Chloro charis emiliae.*"

As noted, the male of *Meruli seebohmi*, the 'Golden-beaked Blackbird' of Burbidge, was seen distantly at about 10,500 feet, but not obtained; though a female was secured at a lower level.

APPENDIX V

DISTRIBUTION OF BUTTERFLIES IN BRITISH NORTH BORNEO

(Made for June and July)

(A Reference Letter—A, B, C, etc.—is noted against each species in the *Index for Butterflies, and in the Supplementary List of Butterflies* overleaf.)

A. Benoni. Sea Shore.
B. Miri (Sarawak).   150 feet.
C. Labuan Island.   150 feet
D. Kotabalud.   300 feet.
E. Kotabalud to Tamu Darat.   300–500 feet.
F. Rayoh (Tenom Gorge).   400 feet.
G. Tenom.   600 feet.
H. Kotabalud to Kabayau.   300–800 feet.
I. Kotabalud to Kaung.   300–1100 feet.
J. Kotabalud to Dallas.   300–2500 feet.
K. Tamu Darat to Kabayau.   500–800 feet.
L. Tamu Darat to Kaung.   500–1100 feet.
M. Tamu Darat to Dallas.   500–2500 feet.
N. Tamu Darat to Bundu Tuan. 500–4900 feet.
O. Kabayau to Kaung.   800–1100 feet.
P. Kabayau to Dallas.   800–2500 feet.
Q. Bundu Tuan.   3300 feet.
       (Actual Base of Mount Kinabalu)

R. Tamu Darat to Tinompok. 500–4900 feet.
S. Kaung to Dallas.   1100–2500 feet.
T. Kaung to Tinompok.   1100–4900 feet.
U. Kaung to Bundu Tuan.   1100–4900 feet.
V. Dallas to Tinompok.   2500–4900 feet.
      (Actual Base of Mount Kinabalu)

W. Dallas to Bundu Tuan.   2500–4900 feet.
(Actual Base of Mount Kinabalu)

X. Everywhere up to   800 feet.
Y. Everywhere up to   4900 feet.
Z. Kamburonga.   7000 feet.
(Upper slopes of Kinabalu)

Note:—The distributions given are those which were observed, and must be regarded as only approximate.

APPENDIX VI

Supplementary list of butterflies obtained and identified in British North Borneo, and not mentioned either in the text ·or in the Index for butterflies

(Reference letters, A, B, C, etc., refer to the table of 'Distribution of Butterflies.' Appendix V.)

Amathusia phidippus (close to) .... (A)
Cirrochroa fasciata fasciata .... (K)
Cyrestis nivea nivalis .... (Q)
Danais aglyia .... (M)
Delias hyparete simplex .... (D)
Dolleschallia bisaltide .... (I)
Elymnias lutescens .... (O)
Elymnias nesaea hyperedes .... (A)
Euthalia dysphania transducta .... (O)
Hesperidae tagiades atticus .... (L)
Hylodes curanea (moth) .... (K)
Idiopsis daos ....
Lithosidae, asota egens indica (moth) .... (O)
Lycaenid, eoxylides tharis .... (K)
Melanitis ismene .... (O)
Neptis hylas mamaja .... (J & Q)
Neorina lowi .... (A)
Papilio damolion .... (J & G)
Papilo memnon isaka .... (A)
Precis iphita horsfieldi.. (F & H)
Terias hecabe (including a white female) .... (Y)